Geographies of Meat

With the ever-rising demand for meat around the world, the production of meat has changed dramatically in the past few decades. What has brought about the increasing popularity and attendant normalisation of factory farms across many parts of the world? What are some of the ways to resist such broad convergences in meat production and how successful are they?

This book locates the answers to these questions at the intersection between the culture, science and political economy of meat production and consumption. It details how and why techniques of production have spread across the world, albeit in a spatially uneven way. It argues that the modern meat production and consumption sphere is the outcome of a complex matrix of cultural politics, economics and technological faith. Drawing from examples across the world (including America, Europe and Asia), the tensions and repercussions of meat production and consumption are also analysed.

From a geographical perspective, food animals have been given considerably less attention compared with wild animals or pets. This book, framed conceptually by critical animal studies, governmentality and commodification, is a theoretically driven and empirically rich study that advances the study of food animals in geography as well as in the wider social sciences.

Harvey Neo is an associate professor at the Department of Geography, National University of Singapore. His research interests include the political economy of meat, green urban development and geographies of food. He is an at-large board member of the Animal Geography Specialty Group at the Association of American Geographers, and editor of *Geoforum* and associate editor of *Regional Studies, Regional Science*.

Jody Emel is a professor at the Graduate School of Geography, Clark University, USA. Her research interests include animal geographies, political economy of mining and water resources. Her current research focuses on the political ecology of factory farming. She teaches courses in natural resource development, feminist theory and nature, hydrology, and the relationship between economy and environment.

Critical Food Studies

Series editor: Michael K. Goodman, University of Reading, UK

The study of food has seldom been more pressing or prescient. From the intensifying globalization of food, a world-wide food crisis and the continuing inequalities of its production and consumption, to food's exploding media presence, and its growing re-connections to places and people through 'alternative food movements', this series promotes critical explorations of contemporary food cultures and politics. Building on previous but disparate scholarship, its overall aims are to develop innovative and theoretical lenses and empirical material in order to contribute to – but also begin to more fully delineate – the confines and confluences of an agenda of critical food research and writing.

Of particular concern are original theoretical and empirical treatments of the materialisations of food politics, meanings and representations, the shifting political economies and ecologies of food production and consumption and the growing transgressions between alternative and corporatist food networks.

For a full list of titles in this series, please visit https://www.routledge.com/Critical-Food-Studies/book-series/CFS

Confronting Hunger in the USA
Searching for Community
Empowerment and Food Security
in Food Access Programs
Adam M. Pine

Geographies of Meat
Politics, Economy and Culture
Harvey Neo and Jody Emel

Forthcoming

Children, Nature and Food
Organising Eating in School
Mara Miele and Monica Truninger

Hunger and Postcolonial Writing
Muzna Rahman

Practising Empowerment
Wine, Ethics and Power in
Post-Apartheid South Africa
Agatha Herman

**Taste, Waste and the New
Materiality of Food**
Bethaney Turner

Geographies of Meat
Politics, Economy and Culture

Harvey Neo and Jody Emel

LONDON AND NEW YORK

First published 2017
by Routledge

2 Park Square, Milton Park, Abingdon, Oxfordshire OX14 4RN
52 Vanderbilt Avenue, New York, NY 10017

Routledge is an imprint of the Taylor & Francis Group, an informa business

First issued in paperback 2018

British Library Cataloguing in Publication Data
A catalogue record for this book is available from the British Library

Library of Congress Cataloging in Publication Data
Names: Neo, Harvey, author. | Emel, Jody, author.
Title: Geographies of meat : politics, economy and culture / Harvey Neo
and Jody Emel.
Description: Abingdon, Oxon ; New York, NY : Routledge, 2017. | Series:
Critical food studies | Includes bibliographical references and index.
Identifiers: LCCN 2016046617| ISBN 9781409440338 (hardback) |
ISBN 9781315584386 (ebook)
Subjects: LCSH: Meat industry and trade--Political aspects. | Animal
industry--Political aspects. | Animals--Economic aspects. | Biopolitics.
Classification: LCC HD9410.5 .N46 2017 | DDC 338.1/76--dc23
LC record available at https://lccn.loc.gov/2016046617

ISBN: 978-1-4094-4033-8 (hbk)
ISBN: 978-0-367-13881-3 (pbk)

Typeset in Times New Roman
by Taylor & Francis Books

To Samuel Keith Lenhardt (1986–2015) and the Neo family

Contents

List of Illustrations   x
Acknowledgements   xi
Abbreviations and Acronyms   xiii

1   Introduction   1
Preamble 1
Commodification 4
Political Economy 5
Biopolitics and Governmentality 7
Critical Animal Geographies and Justice 8
Conclusion 11

**2   The Political Economy of Meat: Global Trends and Local
     Tensions**   13
Introduction 13
Food Regimes 14
*Political Economy of Food and the Complicit Politics of
     Public–Private Sectors 16*
Political Economy, Institutions and Historical Contingency 18
Contract Farming and Corporate Social Responsibility 20
*Normalising Industrial Contract Farming: Case Study of
     Poland 24*
Tensions and Perceptions in Polish Pig Production 27
*When Culture Meets Political Economy: The Contrary Case
     Study of Malaysia 32*
Conclusion 40

3   Science, Technology and the Commodification of Food Animals   41
Introduction 41
Biopower, Biopolitics and Chained Commodities 44
The Political Economy of Animal Science 46

A Brief History of Animal Science 49
Fast and Unnatural Commodity Production 52
'Euthanasia' Experiments and Rendered Commodities 55
Conclusion 61

4   **The Global Meat Factory and the Environment**   64
Introduction 64
The Intensification of the Global Livestock Industry 65
The Ecological Impacts of Livestock Production and
  Consumption 67
Labour Issues and Workers' Health 71
The Political Economy of Environmental Impact Assessment:
  Hoofprint Analysis 76
Who Is Paying Attention to These Data? 78
The Biopolitics of Creating the Model Animal: Ramifications and
  Mitigation 80
Conclusion 82

5   **The Thanatopolitics of Industrialised Animal Life and Death**   83
Introduction 83
What Are CAFOs? 84
The Biopolitics of the Animal-Industrial Complex 86
The Theory and Science of Animal 'Welfare' 87
'There is no word for "pig" in the Lakota language': Case Study of
  South Dakota 91
Animal Welfare in Slaughterhouses 95
Transnational Politics of Food Animal Transportation 98
Conclusion 106

6   **On Not Eating Meat: Vegetarianism, Science and Advocacy**   107
Introduction 107
The Limits of Organic Livestock Production 108
Vegetarianism and Social Action 116
Framing Vegetarianism in East Asia 120
Cultured Meat and Meat Analogues: Resistance or
  Commodification Redux? 128
Conclusion 133

7   **Conclusions**   134
Introduction 134

*Softly-Softly and the Resilient Governmentality of the Meat
    Complex 135*
*Geographies of Meat: The Missing Pieces and Future
    Prospects 137*
Conclusion 142

References   144
Index   169

Illustrations

Figures

1.1	Conceptual framework	3
2.1	Typical pig farm in Turostowo, Poland	28
2.2	Typical storage facility for animal feed in Turostowo, Poland	28
2.3	Distribution of Agri Plus and Poldanor farms and processing plants	29
2.4	Nursery pigs at a pig farm, Malaysia	34
2.5	Weaning piglets at a pig farm, Malaysia. Note the restricted space for the sow	35
5.1	A concentrated animal feed operation in South Dakota, USA. Note the opaqueness of the facility	92
6.1	Jinghong and Jinuo Mountain, Xishuangbanna, China	111
6.2	Entrance of the holding farm for small-ear pigs in Jinghong, China	112
6.3	Tagged pigs in the holding farm for small-ear pigs in Jinghong, China	112
6.4	Specialist butcher's shop in Jinghong, China. The two columns of words in green read: 'Safe meat, organically raised; strengthen your spirit, return to nature.'	113
6.5	Banana tree trunks used as (organic) feed for small-ear pigs	114

Table

6.1	Excerpts from interviews, focusing on animal welfare	125

Boxes

2.1	Farmers' perceptions of the contract grower programme	30
5.1	Sample guidelines for Islamic slaughter	104

Acknowledgements

This modest book would not have been possible without the networks of academics, students, farmers and activists that nurture and provoke our myriad ideas about the geographies of meat. The encouragement of the book series editor, Mike Goodman, is also duly acknowledged. We also record our thanks to Gareth Richards of Impress Creative and Editorial who provided us with meticulous and exemplary copy-editing support.

Harvey would like to thank the geography students at the National University of Singapore (NUS), especially those who have read *Nature and Society*, who have constantly motivated him to look further and deeper in the field of animal geographies. The Department of Geography at NUS has been and continues to be a nurturing and collegial place to be a teacher-scholar. Deep appreciation goes to members of the department's Politics, Economies and Space Research Group who read various parts of the book and never failed to be critical yet encouraging. Special thanks go to: C.P. Pow for his years of friendship which have made Harvey's academic journey ever more enjoyable and bearable; Woon Chih-Yuan for cheering him up when the chips are down; his graduate students – Chua Chengying, Piseth Keo, Guanie Lim, Pamela Teo and Shaun Teo – for their understanding when work on the book sometimes took his attention away from them; the Animal Geography Specialty Group at the American Association of Geographers for being the light bearer for the sub-discipline. Harvey has also benefited much from peers who have researched on aspects of meat animals, such as Alice Hovorka, Mara Miele, Ian MacLachlan and Julie Urbanik, to name but a few. Last, but certainly not least, Harvey would like to thank Jody Emel, his co-author, who has been his mentor and friend all these years.

Jody would like to thank all of the students in the Animal Geographies and Feminism, Nature and Culture courses at Clark University who have pushed her thinking into new arenas. Son Ca Lâm, Catherine Jampel, Ilanah Tavares, Alex Cohen, Leslie Wyrtzen, Padini Nirmal and Alida Cantor are especially important forces for advancing decolonial, feminist, non-anthropocentric thought. A great debt is also owed to the critical animal studies people like Krithika Srinivasan, Julie Urbanik, Connie Johnston, Jennifer Wolch, Henry Buller, Katie Gillespie, Rosemary Collard and Mara Miele for providing a strong and growing foundation for this kind of research and writing. Thanks also to the animal voice activists who work for nothing, shovel manure, gather signatures for petitions, organise protests, go undercover and fight for beings whose lives matter. And for always offering support and encouragement, Jody thanks her family: Julia, Janet, Ted, Jackie and Sam. This work is dedicated to Sam, who bore the sometimes unbearable pain of this world and still made us laugh.

Harvey Neo and Jody Emel
June 2016

Abbreviations and Acronyms

AVMA	American Veterinary Medical Association
BSE	bovine spongiform encephalopathy
CAFO	concentrated (confined) animal feed operation
CO_2	carbon dioxide
CSR	corporate social responsibility
DMG	N,N-dimethylglycine
DNA	deoxyribonucleic acid
DVS	Department of Veterinary Services
ECoG	electrocorticography
EEG	electroencephalograph
FAO	Food and Agriculture Organization
FAWC	Farm Animal Welfare Council
FDA	Food and Drug Administration (United States)
FDI	foreign direct investment
FPRF	Fats and Proteins Research Foundation
FSIS	Food Safety and Inspection Service
GHG	greenhouse gas
HeV	Hendra virus
IGF-1	insulin-like growth factor
ISAE	International Society for Applied Ethology
LiveCorp	Australian Livestock Export Corporation
MOET	multiple ovulation and embryo transfer
MRSA	methicillin-resistant *Staphylococcus aureus*
OIE	Office International des Epizooties (renamed Organisation Mondiale de la Santé Animale)
PAS	Parti Islam Se-Malaysia
PETA	People for the Ethical Treatment of Animals
PSS	porcine stress symptom
R&D	research and development
rBST	recombinant bovine somatotropin
SNP	single nucleotide polymorphism
UMNO	United Malays National Organisation

UNFCCC	United Nations Framework Convention on Climate Change
USA	United States of America
USDA	United States Department of Agriculture
VSS	Vegetarian Society Singapore

1 Introduction

Preamble

In August 2013 the world witnessed the first taste test of in vitro meat, developed by researchers at Maastricht University in the Netherlands and funded by the Google co-founder Sergey Brin at a cost of more than $300,000. The 140-gram saffron-flavoured meat patty, mixed with bread-crumbs and dyed in beetroot juice (because the actual colour of the in vitro meat is an unappetising pale yellow) was described by the taste testers as 'close to meat' and like an 'animal protein cake' (Bloomberg, 2013). The possibility of laboratory-made meat is but the most recent manifestation in the ever-changing nature of meat. Producing meat in such a radical, seemingly denaturalised manner marks the third turning point in humans' relationship with food animals. The first critical turning point was the domestication of wild animals for sustenance and subsistence, as opposed to hunting for them (Caras, 2002). It would be thousands of years after this – in the form of confined animal feeding operations or factory farms – that the next distinct phase of the human–food animal relationship became apparent. The detachment of animals from humans and 'nature', as well as their progressively intensified commodification, arguably comes to its most extreme conclusion with the introduction of synthetic meat.

However, it is clear we are still very much entrenched in a world of intensified livestock farming and the possibility of a market centred on synthetic meat remains a pipe dream for the foreseeable future. The current model of inten-sified livestock farming is one whose negative impacts on the environment, animals and people have been well documented. While we will address these impacts, this book does not aim to merely rehash well-rehearsed arguments *against* factory farming. Rather, our first major goal is to describe the spread and variability of this mode of farming. Second, we explain how and why this form of farming is being progressively replicated in diverse places that hitherto have not fully embraced it. Third, we discuss the limits and potential of resisting the spread of intensified farming in all of its variegated guises, as well as the possibility of offering alternatives. The analysis employs a triad of conceptual frameworks: political economy, biopolitics and critical animal geographies. Clearly these frameworks do not operate in a social vacuum; hence we valorise the broader ideas of culture, technology and values to complete the conceptual and empirical narrative of a dynamic geography of meat. To elaborate briefly for now: the spatial differences in culture (broadly defined), technological competence and varying ethical valuation of food animals all work to influence the receptivity as well as the options of alternatives against intensified livestock production. The focus on the spread of intensified farming is timely, particularly when the livestock industry in general has witnessed dramatic changes in recent decades, and is simultaneously under pressure from climate activists, animal welfare and rights activists, as well as food quality consumer movements (to mention just a few of those critiquing the industry). Yet, in spite of these persistent and trenchant criticisms and some level of institutional changes in response to them – at least in the European Union (EU) and some other so-called 'developed' economies – we are still witnessing the introduction of homogenising, intensified meat production practices to many developing regions. How have such practices been packaged and sold to these places? How successful have such exports been and what are the kinds of resistance against such a spread? In pursuing these and other questions, our ultimate goal is to evaluate the *inevitability* of intensified farming.

While we posit the continuing spread of this model of farming, the production and consumption of meat have always been geographically uneven. Indeed, not all types of meat animals are the subject of intensification processes – a testament to how the taste for meat varies significantly from one place to another. Moreover, besides taste (which is both a driver and reflection of demand), the physiology of certain meat animals, like goats, might be (erstwhile) ill-suited for intensification. In contrast to goats, animals like cattle, pigs and poultry are the popular targets for intensification processes. They also happen to be the three most globally popular meat animals – a fact that is far from coincidental.

Indeed the complex relationship between production and consumption is such that each reinforces the other, whereby high demand drives high supply through intensification which in turn creates even higher demand. As we will show, this is a vicious cycle of supply and demand that ultimately produces rising negative externalities in people, places and animals. In some cases, the ever cheaper and larger supply of meat can reshape the demand for particular kinds of meat in a given locale. In such a way the taste and demand for particular kinds of meat, often underpinned by erstwhile rigid cultural and social constructions of food animals, can change quite dramatically in a relatively short span of time. However, the increased supply cannot happen without technological interventions in the rearing, processing and transportation of meat. Such often brutal interventions (from the perspectives of food animals) in turn must have the complicity of an amoral and/or unaware consumer body to occur. This suggests that to disrupt and rupture this vicious cycle, the consumer (or more specifically, changing the consumer's values) is likely to play a critical role.

The above paragraph sketches in broad strokes the importance of economy, culture, technology and values that are imperative to the pace of transformation in the global livestock industry towards further intensification, as well as the likely paths of resistance. Teasing out the multifarious strands in this 'meat narrative' is not new. Even the earliest works on geographies of meat acknowledged its complexity and ranged widely in their focus, running the gamut from 'mast feeding' (Shaw, 1940) to international trade (Langdon, 1945) to food security (Hilliard, 1969). We will not attempt to cover all possible themes in this book largely because we aim to focus on issues that relate, directly or indirectly, to the intensification of livestock farming.

The overall framing of the book is illustrated in Figure 1.1. Recalling we posit that there is no one fixed definition of intensified farming (or factory farming), we instead suggest 'commodification' as the defining characteristic of intensified farming where the differences in the extent of intensification are directly linked to the depth of commodification. We argue that the commodification of food animals is a direct consequence and acquiescence of several other social and political economic developments. These include an evolving *biopolitics* of food animals and a broader *political economy* of livestock which aim to modernise and extract greater economic capital out of food animals. In addition, drawing on insights from *critical animal geographies*, we argue that the changing sociospatial relationships between humans and non-human food animals have contributed to the commodification process. These relationships are underpinned by a cultural and ethical blindness that obscures and devalues the agency of food animals as sentient beings. To be more precise, making use of the gamut of literature on food justice (notwithstanding the fact that it pays relatively uneven attention to the ethics of the meat industry),

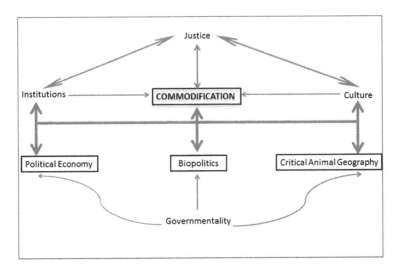

Figure 1.1 Conceptual framework

we excavate the question of justice in the commodification process. Finally, specific institutions (such as government planning agencies, global developmental organisations, agrofood conglomerates and the animal science industry) deepen the commodification of food animals through an overarching governmentality whose ultimate goal is to normalise particular treatments (that is, commodification) of food animals. These series of arguments will be illustrated via a variety of empirical examples, drawn from both the developed as well as developing countries.

In Chapter 2 we outline the contours of a political economy of meat to show how institutions and conglomerates work to transform the production models of pig farming in Poland and Malaysia. Chapter 3 focuses on technology and biopolitics to show how they have become central tools in dramatically altering both the mode of production of food animals as well as our very perceptions of food animals as sentient beings. In Chapter 4 we discuss how, with the increasing awareness of its negative environmental impacts, attempts have been made to neutralise environmentally rooted objections to intensified livestock farming. Extending some of the points developed in the preceding chapter, Chapter 4 looks at how the question of 'environment' (broadly defined) is pivotal both in explaining the spread and resistance of intensified farming. Adding to the argument against the consumption of meat, particularly when produced from a highly intensified setting, there is a need to further valorise the negative externalities of such a production model. In Chapter 5 we turn to the multiple considerations and nuances in animal welfare. How can more knowledge about the ways animal welfare has been severely compromised be employed against intensified farming? Or how might proponents of such a model fend off and deflect these criticisms?

In the penultimate chapter, we consider what is left for the conscionable and the activists against this tidal spread of commodification/intensification of animals. One strategy is to disrupt, through concerted activism, the normalisation of factory farming in general and the consumption of meat specifically. This strategy builds upon our discussion in Chapters 4 and 5. Another strategy is to offer meaningful substitutes to the consumption of meat. Both these strategies are predicated upon a widespread ethical change in human–food animal relations which recognises the latter's sentience, as well as a transformed perspective towards the nature/environment–society relationship. In developing these arguments, we have ranged broadly in the case studies that we draw upon, from the Americas to Europe to Asia and Australia. Having outlined the organisation of the book, the rest of this introductory chapter briefly engages the question of commodification of food animals as well as the three core concepts that underpin it (see Figure 1.1).

Commodification

At the heart of intensified livestock farming, and its more extreme nomenclature like factory farms or confined animal feeding operations, lies the singular drive towards commodification. We find the idea of commodification particularly useful in understanding the variability in intensified farming. To be sure, understood plainly, intensification is a desire for heightened productivity, and it is a goal that is inherent in any sector of the capitalistic market economy, not just in farming. Here, we adopt a broad Marxist critique and understanding of the commodification of living things (including humans and non-human animals) where commodification, at its simplest, reduces and economises the myriad values of an entity into predominantly monetary, anthropocentric ones (Buller and Roe, 2013). In so doing, any social relations that might have existed between food animals and the wider society are obliterated. In other words, critiquing commodification brings to light social relations that might be hidden from view in the more economistic and mechanical term of intensification. Recognising the social relations that exist between food animals and consumers is critical to valorise the immoralities of the contemporary meat production complex.

We highlight, at various points of the book, the ramifications of commodifying animals (as food) and how solutions to address some of these negative ramifications are simultaneously constrained by such commodification, even as they attempt to reframe meat animals as fleshy, sentient beings. Specifically, we are interested in how (extreme) commodification of food animals has come to be normalised and accepted such that the range of resistance against it is still arguably ineffectual. We briefly answer the latter question by introducing the concepts of political economy, biopolitics and critical animal geographies.

Political Economy

A neoliberal political economy thrives by producing and selling ever more goods. Food in general, including food animals, extends beyond being mere consumerist products as it has significant geopolitical and security ramifications. For example, Raymond F. Hopkins and Donald J. Puchala (1978) noted some time ago the broader links between food production and international development. Among other things, their focus flags the issue of food security, which emerged as a critical concern in the 1970s (Atkins and Bowler, 2001, p.154) and has seen a renaissance of sorts in recent years (Lang and Barling, 2012; Veeck, 2013). In a recent article, *The Economist* (2014) highlights the geopolitical security significance that China places on pork such that the central government maintains the world's only 'pork reserves' which release pork into the market in times of scarcity and high prices.

Ostensibly set in less-developed countries, such works articulate the complex causal relationships between food provision, politics and (the imperfections of) global/national economic development (Scott, 1976). Key issues investigated thus include the social/political roots (as opposed to, say, strictly environmental causes) of famine or underproduction of food (Nicholson and Esseks, 1978; Blaikie, 1985) and the role of global developmental institutions in alleviating/ aggravating food production (Weiss and Jordan, 1976; Johnson, 1978). The driving force behind these studies revolves around finding the ways to achieve security and development through increased production. However, the empirical focus of most, if not all, of these works is on staple food. With global meat production tripling over the last four decades and increasing 20 per cent in just the last 10 years, it is imperative to turn our attention to meat. Almost everywhere consumers are consuming more meat, with the industrial countries consuming nearly twice the amount of meat than developing countries (Worldwatch Institute, 2013), although in terms of growth of demand, the latter countries lead unambiguously.

As evinced by its geopolitical and food security dimensions, the food economy is first and foremost political. A political economic perspective sees the market economy as inherently political. The production and consumption of food animals is hence always political. For example, policies surrounding the livestock industry do not occur in an apolitical environment. Indeed, broader issues like farm subsidies, consolidation and mergers in the meat industry, the development of and investments in animal science technology and not forgetting opposition towards particular forms of food animal production and consumption are thoroughly political.

The political economy of meat (especially the top three popular meats) in many developed countries is marked by increasing consolidation and upscaling of production. Tony Weis (2013) goes so far as to argue that, among other reasons, it is largely this development that has spurred insatiable and wanton appetites for meat through the production of cheaper meats. In this regard, researchers have found that in large, high-capacity, high-technology pig farms (for example, 15,000 market pigs per year) production costs were 25 per cent lower than a small, low-technology farm that produces 2,000 pigs per year (Barkema, Drabenstott and Welch, 1991). While the trend towards larger farms is a global phenomenon, the more subtle yet no less significant change has been how farms (both big and small) have been increasingly integrated into larger conglomerates (or 'meat processors') through a variety of ways, for example as contract farmers to the larger companies or becoming their sole suppliers at various stages of the meat production processes. In 2012 seven of the largest meatpacking companies were located in the Americas (Tyson Foods, Cargill, Smithfield Foods and Hormel Foods in the United States and JBS, BRF and Marfrig in Brazil); two in Europe (Vion and Danish Crown AmbA); and one in Asia (Nippon Meat Packers in Japan). In terms of annual food sales, the top three companies (JBS, Tyson Foods and Cargill) each made more than $30 billion in 2012. The remaining seven companies achieved sales of from $8.2 to $14.9 billion (Heinrich Böll Foundation, 2014, p.13). These are clearly massive conglomerates. The size and scale of such meat producers and the amount of meat that they produce were unthinkable barely half a century ago. Global meat production in 1961 was only 70 million tons; in 2012 this figure was estimated to be slightly over 300 million tons (Nierenberg and Reynolds, 2012). Another study indicates that poultry is the most popular meat in the world, with an estimated 58 billion chickens and 2.8 billion ducks slaughtered in 2011; the corresponding figures for pigs and cattle are 1.4 billion and 296 million respectively (Heinrich Böll Foundation, 2014, p.15).

The livestock sector has thus clearly undergone significant restructuring in the past few decades and a particularly fertile line of research is precisely to understand the political economic drivers behind this restructuring (Fagan, 1997; Pritchard, 2000; Whatmore, 2002; McMichael, 2009). It is reasonable to assume that such changes, aimed at increasing profit through increased commodification and intensification, are fuelled by the interplay of changing economies, technologies, policies and shifting consumer demands; all of which are underpinned by changing social norms and expectations. Commodification, however, does not and cannot proceed unfettered.

Increasingly, normative questions about the 'quality' of the products as well as their production processes have become imperative to our understanding of the livestock industry. Put simply, the externalities that result from commodification are becoming central to the political economy of the meat industry. Among other things, a political economy that is propelled by profits is said to have negative repercussions on the culture and livelihoods of people and places. On the other hand, culture (broadly defined) can stymie the unquestioned march of the political economy of livestock intensification in some places (Neo, 2009; see Chapter 2). The centrality of both politics/political economy and sociocultural norms in the geographies of food production must thus be recognised (Pence, 2002; Coff, 2006).

Biopolitics and Governmentality

Political power, however, does not sit static in just one place, for example in the economy, even though as indicated earlier the economy is unquestionably political in nature. In a series of works Michel Foucault (2003, 2009) argued that political power is circulated through and constitutes the very foundation of society. The corpus of Foucault's work is extensive and we focus on his notion of biopolitics and biopower. One can understand biopolitics as the process through which groups of beings are understood, quantified and governed. Biopower, as Adrian J. Bailey (2013, p.204) argues, not only 'disciplines individuals as bodies', it also turns on the 'desire, reflexivity and an affirming of life through the coming together of groups'. Put simply, biopower as exercised through biopolitics moulds behaviours even as it allows for a reaction against such a moulding.

Biopolitics has been applied in an array of empirical and theoretical contexts, ranging from spatial governance (Rose, 2013), climate-induced migration (Reid, 2014), conservation science (Biermann and Mansfield, 2014) to labour geographies (Labben, 2014). As Chris Wilbert (2007, p.103) puts it, 'biopolitics involves an incorporation of bodies and behaviour into webs of rules and codes of conduct, and an extension of institutions into more and more aspects of everyday life, especially for matters of health'. What unites these diverse studies is the view that biopolitics (with its attendant 'techniques of power') is essentially a series of deliberate tactics that aim to converge citizen-subjects (and we do explicitly include non-human animal subjects here) towards a broader social norm – the latter which is dictated, sometimes vaguely and imperceptibly, by the powers that be. Social norm here can refer to accepted actions with regard to a whole repertoire of behaviours, including, for the purpose of our book, consumption patterns. There is thus a distinct governmentality at work here in the *normalisation* and biopolitics of consumption. As Nancy Ettlinger (2011, p.538) argues, 'governmentality offers an analytical framework that is especially useful towards connecting abstract societal discourse with everyday material practices'.

The role of governmentality and its concomitant formal regulation in shaping and mediating the economies and cultures of food production are critical in our understanding of the spread and resistance of commodification of food animals. However, governmentality is a process which is often abetted by other concepts like institutionalism (discussed in greater detail in Chapter 2). Our argument is grounded in the belief that institutions (encompassing state, private and non-governmental) help shape the ways in which food animals are produced and consumed. Such an institutional perspective, broadly conceived, also recognises the importance of culture, history and politics. We invigorate this economically driven institutional approach with the concept of governmentality (Huxley, 2007). To put it simply, we argue that the institutionalisation and governmentality of food animals are essentially political contestations over the production and consumption of meat. Such contestations are fundamental struggles over the commodification and meaning of animal bodies (as well as the resistance to such commodification) – a form of biopolitics. Biopolitics then is a specific subset and outcome of governmentality.

In other words, the concept of governmentality allows us to see how formal rules and regulations as well as sociocultural beliefs (which arose from specific institutions) are normalised and subsequently internalised by the subjects who are being governed. It further lets one trace the ways in which these subjects have been transformed subtly, or, in some cases, fundamentally, by such governance. As Mitchell Dean (2010, p.20) puts it, governmentality is the purposeful employment of knowledges and technologies which alter the 'choices, desires, aspirations, needs, wants and lifestyle of individuals and groups'. We wish also to claim here that such alterations can be seen not only in human subjects (in the form of consumers, farmers or even economic structures) but also in food animals as well. Indeed, animals in general have been a productive line of enquiry in geography since the 2000s.

Critical Animal Geographies and Justice

Contemporary approaches to understanding human–animal relationships demonstrate a theoretical and conceptual break from its origins in zoogeographies, rooted in the physical and biological sciences, and early cultural geographies of animals, most associated with Sauerian cultural-geographical approaches to agriculture and domestication (Wilbert, 2009, pp.122–3). 'New' animal geographies emerged out of geography's own cultural turn and interactions with critical social theory, cultural studies and environmental studies.

These enculturated human geography perspectives led to profound rethinking of culture, subjectivity and nature (Wolch, Griffith and Lassiter, 2002), contributing variously to the development of critical animal geographies. First, there is increasing work by geographers on embodied and performative experiences and calls to refocus on the material and social aspects of human life, translating to attention on both the 'fleshiness' and corporeal aspects of animal bodies (Wilbert, 2009, p.126). Second, post-human decentring of the human subject in the face of rapidly advancing technoscience (Castree and Nash, 2006) necessitated a recognition that the human and non-human spheres could no longer be thought of as exclusive, but are mutually co-constitutive. Humans and non-humans are better thought of as a collective or network of actants (Latour, 2004), thereby opening room for the inclusion of subjective agencies of the non-human in shaping human identity (Fox, 2006; Emel and Neo, 2011). Finally, animal geographies have been shaped by feminist and postcolonial critiques of the treatment of human–non-human and human–animal relations centred on the white, male subject (Anderson, 1997; Emel, 1998), and the unpacking of these relationships and the production of our knowledge(s) of them (Braun and Castree, 1998; Castree and Braun, 2001). More recently, work has also demonstrated how non-human animals are implicated in human sexualities and sexual practices (Brown and Rasmussen, 2010).

The reframing of these strands of thought into animal geographies has been productive. Jennifer Wolch and Jody Emel (1998, p.xiii) demonstrate how 'taking geographical approaches to the animal question, or the issue of human–animal relations, will generate rich and provocative ideas'. These are ideas that could be used to excavate the range of human–animal networks and show how these relations are constituted by, and make a difference to, the spaces and places in which they occur (Philo and Wilbert, 2000, p.5). They focus on the discursive and material practices around where humans place animals, physically and metaphorically. Wolch and Emel (1998) suggest four nested scales with which human–animal relationships can be analysed. In increasing order, they are: between individuals, in borderland communities, the political economy of animal bodies, and ethical and moral landscapes. In doing so, they explore the contingency of animal–place orderings and how they are specific to different spatial-temporal and sociocultural contexts; how these orderings are applied differentially by species; and how animals themselves are implicated in both the orderings and their disruption (see also Hovorka, 2006). Such orderings not only have profound implications in the way consumerist society values food animals, they also impact on the way we value and view other actors in the production chain of food animals. At the risk of simplifying a complex body of work, the diverse range of works by critical animal geographers is united by the goal of seeking justice for (and with) animals (Buller, 2013b).

The political economy of commodification of food animals and its attendant governmentality aim to diminish such an insight from critical animal geographies, in particular the way human–animal relationships have been reconceptualised as co-constitutive and grounded in ethical justice. It does so by pursuing a narrow and reductive line of enquiry predicated upon a production–consumption dichotomy. For example, this line of enquiry might investigate whether changes (both qualitative and quantitative) in demand and supply are producer-driven or consumer-driven. As many scholars of agrifood have argued, such a line of enquiry is asking the wrong question at best, and ontologically misleading at its worst, because it assumes that production and consumption can be neatly and realistically separated (Goodman, 2002; Holloway, et al., 2007; Ritzer and Jurgenson, 2010). This is hardly a novel argument. For example, Marx's (1867 [1977]) theory of social relations of production is essentially a complex argument which, among other things, suggests that production and consumption are intricately related social processes. Analytically, production and consumption have a reiterative relationship and cannot be seen as two dichotomised spheres.

However, particular actors in the food animal networks are insistent on precisely such a separation between production and consumption. This is because such an insistence not only obscures the social relations between consumers of meat and food animals, it also renders invisible the workers who work with food animals and produce meat. Drawing on both Marxian and Foucauldian perspectives, Lourdes Gouveia and Arunas Juska (2002, p.372) show how 'production and consumption become separated in the elaboration of contemporary agrofood systems' and that such 'fictional separation is an artifact of power and socio-cultural, as well as ideological, construction'. This is an ideological construction, they argue, which ultimately subjugates workers and consumers as well as food animals, often to the benefit of private business interests. Hence, a separation of production and consumption processes is not only a prerequisite in the commodification process of animals, it also elides the need for justice for these animals and other human actors.

Suffice to say, drawing from critical animal geographies, we are against the separation of production and consumption of food animals because the alienation of animals as mere consumer products has resulted in them being reproduced as wanton exchange values (for ever-increasing profits) rather than reasonable use values (for sustenance). This in turn blinds consumers to the unethical production processes of food animals and results in these animals being unjustly hindered and thwarted in their fundamental needs and capacities in a human-mediated and dominated world. Taking the issue of justice of (and for) animals (as well as the human actors in the production of meat) seriously is one key strategy to denormalise commodification. For example, drawing on the example of organic food consumption, Emma J. Roe (2006) presents the concept of 'embodied practices' as a way to shift focus from the usual producer/consumer dichotomy to the bodies of humans and non-humans so as to better understand the complex ethical relationships between food consumption and food production.

To reiterate, the dichotomised formulation of the food animals system into production and consumer spheres, in its reductionism, produces an inexact, if not outright false, reality. More importantly, such a reductionist ideology invariably leads to the continued marginalisation of animals and peoples by blinding the consumer at large to the injustice of the food animal system. However, having said that, we too recognise the heuristic convenience and utility in drawing on such a dichotomy, not least because it sensitises one to the fact that the power relations undergirding the meat networks are often in favour of private and state interests (that is, certain actors in the 'production' realm). In other words, a relational perspective of meat systems which collapses the production and consumer dichotomy must concomitantly recognise the unequal power relations between the 'production end' and the 'consumer end' of the relationship.

Conclusion

Meat functions not merely as a source of sustenance. As with other forms of consumption, it is also a means of identity building (Jackson, 1987; Jabs, Sobal and Devine, 2000; Guthman, 2003; Buerkle, 2009; Tserendejid, et al., 2013; Bell and Neill, 2014). In this book, we view meat as more than functional. This is in itself unsurprising for, as noted earlier, the production of food (including meat) is essentially one exemplification of social relations at work. In that sense we are particularly wary of claims that assert only the functionality of food animals, especially when framed in terms of them being quantifiable, profit-generating objects or commodities. Arguably, meat remains one of the most ubiquitous yet taken for granted and neglected material manifestations of social culture. More than ever, pertinent and complex interconnections between culture, values, development, politics and food/meat production and consumption remain to be teased out (see, for example, Murray, 1998; Robbins, 1998; Atkins and Bowler, 2001; Counihan and Van Esterik, 2013). We seek to illuminate these interconnections in this book, through valorising both the function and the form of commodified meat, and their tensions therein.

This introductory chapter lays out the main direction that the book will take and sets the general context that drives the study. For the latter, while we recognise the broad trend of the supply of increasingly homogenised and commodified food animals that caters to a rising demand, we wish to also explore the countervailing evidence that defies this. In so doing, we also expose the failings of the hegemonic construction of food animals as a commodity. These are failings which amplify the diminishing welfare of peoples and animals associated with the meat industry.

Yet, such commodification does not come naturally nor is it inevitable, for there remain space and place for alternative forms of livestock production that are not driven by commodification or intensification. Through the concepts of political economy, biopolitics and critical animal geographies, via governmentality, we attempt to show how knowledges and practices surrounding the meat industry are (re)produced to make commodified food animals the norm. Against this, we want to highlight the abnormality of the contemporary meat system. We do this by first focusing on the socioeconomic and environmental impacts of intensified meat production. Some of these impacts include the plight of small-scale farmers; the contested transformations of rural communities; the increased health risks and environmental pollution as well as the ethical and just treatment of livestock (for example, the way they are reared and the modification to their genetic make-up). Second, we show how the political economy of meat producing comprises various actors across scales which enact policies and practices that normalise commodification. These concerns and expositions – which tend to portray the corporate livestock producers as the source of many of the contemporary problems faced by the livestock industry – are imperative in any debates over the 'modern' meat industry but ultimately beg more questions. How did such commodification emerge in specific places? What are its ramifications? What can be done to resist such commodification and who are the ones resisting it? To what extent and in what ways would these resistances be considered fruitful or successful? These are the main questions driving this book.

2 The Political Economy of Meat
Global Trends and Local Tensions

Introduction

Nearly 30 years have passed since the geographical journal *Area* published a short research note urging practitioners to redefine agricultural geography as the geography of food (Atkins, 1988). This clarion call was aimed at broadening the scope of agricultural geography as a subdiscipline and suggested several ways of doing so. These included redressing the 'unfortunate neglect by agricultural geographers of the rural development in poor countries' (Atkins, 1988, p.281); exploring the role of food systems in developmental processes; and reviving the geography of diet and nutrition, aimed at shifting the focus on 'producer and production' to 'product and consumer'. In retrospect, P.J. Atkins's disclaimer that his three lines of enquiry were 'but a sample of the many that one could identify' is a gross understatement given the proliferation of empirical and theoretical research on the geographies of food in the past two decades. Not least, food, including meat, production is deeply entwined in sociopolitics and engenders a plethora of economic tools and policies. Many contemporary studies of agrifood aimed exactly to broaden the base of agricultural geography to 'subsume the non-farm elements of the food system' (Atkins, 1988, p.282).

Given our interest in the advent of intensified livestock farming across scales and places, this chapter begins with a discussion of a global perspective of food production systems, drawing on insights from food regime theory. While there is much to be gained from an awareness of such 'globalised' approaches, we highlight their inadequacy in three ways. First, although they rightly valorise the importance of both politics and the economy in shaping and regulating the development of food production, the specific nature or the form in which the political economy of food takes shape is seldom discussed. Second, much of this work views the way food production has developed (at the broadest scale) as almost teleological. In that sense, while it usefully describes some of the broad converging trends at the global scale, it does not necessarily and rigorously investigate the resistance against such trends. Third and relatedly, such a teleological gaze has the unintended consequence of normalising, or worse, sanctioning the very development itself. To address some of these oversights and enliven the discussion, we draw on a case study in Poland that deals

with the flows of foreign direct investment (FDI) and its socioeconomic ramifications. Concomitantly, we highlight the hidden knowledges, power relations
and discursive constructs that surround these issues which ultimately serve to
govern and control subjects (that is, food animals and farmers). We also offer
a contrary case study in Malaysia where local cultural politics of consumption
stymie the advent of global industrial livestock production.

Food Regimes

Rooted in the world system paradigm (Wallerstein, 1974), the main proponents
of the broadly structuralist conception of food regime are the sociologists
Harriet Friedman and Philip McMichael. Food regimes are said to be 'the
rule-governed structure[s] of production and consumption of food on a world
scale' (Friedman, 1993, pp.30–1) and 'are found in the characteristics of large-
scale food production and consumption and their relation to the state system'
(McMichael, 1992, p.344). The focus of food regime research is to explicate
the regulatory frameworks governing food production (and to a lesser extent,
food consumption). Three different food regimes have been elaborated. The
first spanned from the late 1800s until just before the First World War. This is
a global colonialist-based regime 'centered on the export of meats and grains
from settler states to Europe ... in exchange for capital and manufactured
goods' (Pritchard, 1998, p.66).

The second regime can be characterised as a 'surplus regime', spanning
from the period after the Second World War to the early 1970s. Key features
during this period include the intensification of production, new technologies of
production and a regulatory shift towards increased protectionism. The surplus
regime (surplus because of its characteristic overproduction of grains) is largely
the result of the hegemonic US model of 'agriculture as an integrated sector of a
national industrial-capitalist economy' (Talbot, 1995, p.140). Encapsulated
within this second food regime are the tendencies towards the transnationalism
of food production which will disrupt the stability of the regime. Other notable
developments that arguably fall outside the second food regime include:

1   the perennial efforts at reforming the protectionist agricultural policies of
    Europe and the United States;
2   the emergence of 'newly agricultural markets' which can compete in a
    liberalising world market for agricultural products;
3   the growing demand for processed food on one hand and the concern for
    food safety and quality;
4   the growing dependence on transnational companies for seeds and other
    farm inputs (for example, pesticides).(Talbot, 1995, pp.140–1)

That there have been very few recent works that deal with the concept of
food regime specifically suggests that, notwithstanding its useful focus on
broad-scale regulation, the food regime approach in the study of agrifood can

be and should be augmented by other cognate concepts and its empirical gaze suitably broadened. There are several reasons why this has to be so. First, as food regime research is predominantly grounded in the global scale, it tends to overlook meso- or local-level food sectors which might not be significantly integrated to any global system. Moreover, in focusing exclusively on regulatory aspects, it ignores other important dimensions of agrifood. These include, for example, the significance of culture in contradicting broad trends in the production and consumption of food, as will be discussed later in this chapter. In sum, as Prabhu Pingali and Ellen McCullough (2010) argue, the drivers of change in the global livestock industry go beyond trade and capital (the focus of regime theory) to include issues like income growth, urbanisation, technological change and the rise of female employment. We suggest that a significant failing of the food regime approach is its tendency to ameliorate local complexities in telling a global tale of homogenising and hegemonising agrifood production.

In that sense, the usefulness of delineating a 'third food regime' is debatable. W.N. Pritchard (1998, p.64), for example, describes the third food regime as one in which 'strategies for profit capture are built around internationally coordinated flows of production, commodities and money capital' (see also McMichael, 1992). Some food crops have indeed been very much drawn into a globalised network of flows of goods (see Tan, 2000 for a case study of coffee networks in Vietnam), and the global meat complex (Schneider and Sharma, 2014) has in some instances indeed fallen into the contours of the third food regime. However, many other food crops are much less globalised; moreover, as highlighted earlier, trends in agrofood production and consumption are often contingent on local context. As Norman Long and Jan Douwe van der Ploeg (1988, p.37) put it:

> Agricultural development is many-sided, complex and often contradictory in nature. It involves different sets of social forces originating from international, national, regional and local arenas. The interplay of these various forces generates specific forms, directions and rhythms of agricultural change.

In other words, because there might not be one dominant process at work (but several, or none), the notion of 'regime' as often used structurally in food studies invariably reduces all changes to the imperatives of an international political economy (be it colonialism, global regulatory framework or multinational companies) and is hence unnecessarily stifling and limiting. Among other implications, it might simplify key developments in livestock development, assuming the notion of 'intensification' (for example) as uniformly applied and unproblematic. Hence, one must take care not to overstate the globalisation of food production because a more nuanced attention to local specificities is imperative for agrifood studies (even with respect to 'globalised' food crops like cocoa, bananas and coffee). However, we note favourably the strong

defence of the food regime approach mounted by one of its earliest proponents, who argues that 'debates over the composition and significance of a
food regime [such as those briefly outlined above] have been productive in
expanding its analytical reach' (McMichael, 2009, p.140). As far as expanding
its analytical reach is concerned, we nonetheless believe that the political
economy of food approach, in 'bringing down the analysis', complements the
food regime literature by more pointedly focusing on the intimate interactions
of socioeconomic realities between various scales, from the local to the global.

Political Economy of Food and the Complicit Politics of Public–Private Sectors

To be sure, their complementarity should not come as a surprise as the political
economy approach to food is actually closely aligned to food regime theory.
However, unlike the latter, the former covers a broader, more disparate set of
empirical interests and also looks to more diverse sources to explain the drivers behind the development of agrifood industry. For Ben Fine (1994a,
p.520), one of its strongest proponents, the political economy of agrifood

> not only addresses the political economy of agriculture, the issue of
> exchange entitlements, and the role of the state, it also incorporates the
> role of food processors and retailers and seems required to enter that
> postmodern world, or nightmare, in which food takes on the significance
> that we are what we eat.

The political economy of food thus covers a gamut of themes and disciplines
ranging from economics to politics to cultural studies. This attempt to be
comprehensive is laudable but what precisely unites these disparate concerns?
Fine suggests that the conceptual lens of 'system of provision' coherently
maps out the terrain of a political economy of food. System of provision here
refers to 'the chain of activities connecting initial production to final consumption' (Fine, 1994a, pp.520–1), not unlike a 'life cycle' analysis of food
(Weber and Matthews, 2008; de Vries and de Boer, 2010). This approach thus
has clear affinities with the commodity chain theory which has been used
particularly to trace the production and consumption of consumerist goods
like diamonds and gold (Hartwick, 1998). Fine, however, argues that food
systems are distinguished from such non-food products by the 'dependence
of food systems upon a particularly high "organic" content at its extremes
of farm and home, with corresponding implications for economic activity
connecting the two extremes' (Fine, 1994a, p.521; see also Chapter 4). He
envisages analysing different food systems of provision for different foods.
The key issue that needs to be investigated is the relationship between 'the
(re)structuring of the systems of provision, the role of (and distinctions)
between tendencies and trends, and the scope of historical contingency'
(Fine, 1994a, pp.538–9).

There is much to commend in this food systems of provision approach. For example, it recognises the multispatiality and multiscalar nature of food production. It also emphasises the importance context (that is, 'historical contingency') plays in the development of particular food industry. The key point is that the restructuring of food industries is influenced by global trends – a nod to regime theory – but concomitantly mediated to various degrees by local (often contradictory) tendencies. This insight allows one to explain 'anomalies' in terms of differences in cultural and political realities. Others, however, have taken Fine to task for, among other things, his narrow focus on 'rent relation' and 'historical forms of landed property' in order to delineate a political economy of food. There is some truth in this because Fine seems to take rent relation and historical forms of landed property as the main examples of 'historical contingency'.

David Goodman and Michael Redclift (1994, p.551), for example, have argued that Fine's historical contingency is unnecessarily restrictive and fails to 'understand the social production and construction of nature, food and "agri-culture" systems'. For them, a focus on 'nature-society dialectics' is said to better analytically frame 'the complex interactions ... between nature as environment and the political economic and cultural dimensions of food, its production, distribution and consumption under capitalism'. Yet, even though he personally does not dwell on them, Fine's argument can potentially accommodate issues that his critics deem important like 'social agency, contestation and contingency' (Marsden, et al., 1996, p.364) – much like how the food regime framework can conceivably address contemporary developments in the agrifood sector. Hence, his works seem to be harshly read. For Fine (1994b, p.579), 'the reproduction and transformation of the food system potentially occurs integrally along the range of its activities, as analytically addressed, for example, by the notions of appropriationism and substitutionism; the global, the cultural and the state are crucial'. His monograph published in 1998 sought to develop this argument more comprehensively (Fine, 1998).

If anything, the preceding discussion highlights a popular focus in the political economy of food studies that looks at the impacts changing regulatory regimes – at various spatial scales – have on the production, consumption and investments of food processes. It is this focus which highlights the relationship between public and private sectors that we find most useful in explaining (in part) the emergence and normalisation of particular forms of meat production systems. State policies and politics directly impact the nature of food and meat production allowed or encouraged in any given place. For example, the Chinese government has in the past decades adopted a slew of 'pro-pork policies includ[ing] grants, tax incentives, cheap loans for farms and free animal immunisation – all intended to boost intensive pig farming'; through these policies, the Chinese government is estimated to have subsidised pork production by $22 billion in 2012 or $47 per pig (*Economist*, 2014). The intervention of the Chinese government in the pork market reached a peak in 2014 when, through a multibillion-dollar loan by the Central Bank of China, it facilitated the takeover merger of Smithfield by WH Group (previously

known as Shuanghui International Holdings), a domestic Chinese pork processor (Halverson, 2014). Clearly, the Chinese government's policy towards
food security and the consumption of pork is one geared towards both
increasing intensification in China as well as scouring the world for multiple
sources of meat imports. As Mindi Schneider and Shefali Sharma (2014, p.11)
note, the 'dialectic between pork's sociopolitical importance in China and its
mounting externalities will shape Chinese policy towards pork production,
trade and consumer choices in the coming decade'. In China, the complicity
of the public sector with the private sector in promoting a particular kind of
food production and consumption is strikingly clear.

It is not to say that the dictates of the central government will always be
uncontested. Working on Britain, Michelle Harrison, Andrew Flynn and
Terry Marsden (1997) have analysed the tensions in implementing national
food polices and regulation at the local level (see also Lowe, Marsden and
Whatmore, 1994). Changing regulations across scales has caused agrofood
firms to seek opportunities elsewhere or restructure their operations to achieve
greater profits. Bill Pritchard's (2000) study of agrifood conglomerate Nestlé
in Southeast Asia, and Neil Wrigley's (2002) study of the growing dominance
of large food retailers in the United States are representative of such studies.
More relevant for this chapter are those which attribute changing regulatory
regime as partly the result of political economic imperatives. Julie Guthman
(1998), for example, has shown how the political constructions of food and
nature are integral to the formulation of the regulatory regime of organic
farming in California. Becky Mansfield (2003), in her comparative study of
the American and the Vietnamese catfish industries, similarly emphasises the
importance of politics across various scales in the formal and informal regulation of the American catfish industry. In the Malaysian example detailed later
in the chapter, we also see how inconsistent sociohistory and political regulation
disrupt the development of the pig industry.

It is the focus on historical contingencies, the politics of food production/
consumption and the attendant interest in regulatory regimes and issues of
complicit governance that forms the basis of a political-economic perspective on
meat production and consumption. Such a comprehensive political-economic
perspective lends itself well to the concept of governmentality of meat. In what
follows, we first briefly elaborate on the relationship between political economy,
institutions and historical contingency before specifically looking at two
major political-economic strategies: contract farming and corporate social
responsibility (CSR). We argue that these aim to commodify food animals
and assure the success of such commodification by obscuring the negative
repercussions of such a process.

Political Economy, Institutions and Historical Contingency

We use the example of FDI in the Polish pig industry since the 1990s to show
how the former has changed the nature of the local pig industry. While it is

obvious that the relentless pursuit of profits drives meat companies' expansion into overseas markets, clearly many more pertinent questions with respect to such transnational investments remain to be answered. On the one hand, we recognise that national governments actively court such transnational investments; on the other hand, we avoid representing such investments as *necessarily* a kind of top-down 'forced' restructuring that invariably disrupt the local industry and economy (Phelps and Wood, 2006). To that end, we examine the contrast in local responses to transnational capital from two different sources. In focusing on two different streams of FDI to the same region (north and north-west Poland), we offer a more nuanced reading of the changing Polish pig industry that aims to show how intensified production systems and firm strategies are reshaped and received in subtle ways by the specificities and histories of the (post-socialist) 'local'. Concomitantly, we heed the call of Michael Storper (2009, p.18) to investigate 'how geographic and organizational changes in the production systems are altering context, behaviours and choice of development pathways'. In other words, we are using the case study of the Polish pig industry to interrogate more broadly key questions of FDI and governmentality in a post-socialist, transitioning political economy, and how they normalise particular modes of production.

Our analysis is also largely grounded in an institutional economic geography approach that pays close attention to the roles of institutions, history and place-specific political economies in explaining the changes in the pig industry (or in Fine's words, 'historical contingency'). We argue that while there are clear differences between socialist and post-socialist modes of production (as seen through the specific changes within the pig industry in terms of production systems, regulation and social relations), the resulting benefits (or costs) are not easily determined or determinable nor should we discount the lingering effects of the socialist-era political economy.

From an institutional perspective, change is meant to 'reduce uncertainty by providing a structure to everyday life' (North, 1990, p.3). However, in explaining regional disparities, it has also been pointed out that (a region's) 'social infrastructure can help or hinder economic growth' (Cumbers, MacKinnon and McMaster, 2003, p.326). In the past decade or so, many have produced works on institutional economic geography that are both theoretically rooted (Amin, 1999; Jessop, 2001; MacLeod, 2001; Healey, 2006) and empirically focused. The latter include such diverse topics as the semi-conductor industry in north-east England (Dawley, 2007), regional development in British Columbia (Markey, 2005) and FDI in Poland (Hardy, 2006). Despite such a diversity of work, many who write from an institutional geography approach generally accept the idea that the 'institutional regime' (comprising 'institutional environment' and 'institutional arrangements') is key to understanding regional or national economic growth as well as industry-specific development. Institutional environment refers to both the systems of informal conventions, customs and social norms as well as the formal structures of rules and regulations while institutional arrangements, to put it simply, are the material forms of organisations associated with the institutional environment (Martin, 2000). These would include markets, firms, regulatory agencies and so forth.

We argue that it is useful to infuse institutional analyses with a more nuanced understanding of flows of power afforded by the concept of governmentality. In so doing, institutions and governance can be understood as providing opportunities as well as constraints in economic development and policies, along with a keen recognition of the importance of culture, history and politics. In the next section, we look at the implications of contract farming as a mode of production/governance and the importance of CSR in the meat industry. We view both of these as strategies that aim to normalise 'modern' industrial farming in post-socialist Poland, given that such industrial farming is incongruent with the form and history of pig farming in Poland in the socialist era. In brief, the case study of the changing Polish pig industry suggests that receptivity towards the intensification of meat industry (or put another way, its *normalisation*) is not straightforward, immediate or unproblematic. The following three sections are drawn in part from Harvey Neo (2010) and Grzegorz Micek, Neo and Janusz Górecki (2011).

Contract Farming and Corporate Social Responsibility

As mentioned briefly earlier, the livestock industry is continually intensifying through, *inter alia*, 'better' technologies (see Chapter 3 for more in-depth analysis). For transnational FDI from developed countries to less-developed countries, not only would there be a transfer of new technologies, there would also be organisational change in the systems of producing meat. The latter often refers to vertical integration of livestock industry, of which subcontracting (which itself is predicated upon appropriate improvements in production technology) is the primary strategy. Subcontracting is chosen because we see it as one of the more recent, and certainly one of the most dramatic, changes in the political economy of the livestock industry in less-developed places (see Neo, 2010 for a more detailed discussion). The novelty, scale and pace of such changes meant that they require capital *and* the cooperation/receptivity of local governments and farmers.

Although contract farming is pervasive in meat production (for example, virtually all poultry produced in the United States is through contract), it has received considerably less attention compared with the subcontracting of other manufactured products like shoes, fashion or electronics. Peter D. Little and Michael D. Watts (1994, p.5) pointed out early on that the heterogeneity of contract farming is a diverse constellation of crops, actors, production relations and institutional links. Hence, to 'outline a general "theory" of contracting would be foolhardy and ultimately unproductive'. Nonetheless, one can discern two main types of subcontracting relationship in the livestock industry. The first is a marketing contract where the farmer maintains full ownership of the animal that is being bred until the animal is ready to be delivered. In such contracts, the price of the animal at delivery is usually predetermined. In other words, although the farmers typically have a free hand on how the livestock should be raised, they also bear a considerable degree of risk (for example, if an epidemic hits the herd). On the other hand, a production contract sees the principal firm (or, in the case of the livestock industry, the processor) having much more control of the product. In most cases the processors not only own the animals, they would provide the feed, methods of production and veterinary support to the contract farmers. Even the housing facilities of the animals have to be built to the strict specifications of the processors. The contract farmers are, in this sense, virtual paid labourers.

Similar to subcontracting in other sectors, subcontracting in the livestock industry is an organisational innovation that explicitly aims to increase productivity through the use of efficient and easily adoptable production techniques. Yet, because of the nature of its products (for example, the requirement for strict health and safety monitoring and the logistical challenges of moving live animals across distances), subcontracting within the livestock industry is far less internationalised than, say, the electronics industry. In other words, livestock subcontracting networks are often localised in particular regions of a country and cross-boundary subcontracting networks are rare. In less-developed countries, such as Poland, subcontracting in the livestock industry has been largely precipitated by transnational FDI. Nonetheless, these developments require capital and the cooperation/receptivity of local governments and farmers. In this respect, attesting to the complicity of public–private sectors, some countries (for example, Mexico and Kenya) have actually initiated state-led contract farming, with the aid of transnational capital (Little and Watts, 1994, p.8). In China contract farming is the government's 'primary strategy for vertical integration and coordinating rural production along market-based lines' and the Ministry of Agriculture had even established the Office for Vertical Integration of Agriculture in the mid-1990s (Schneider and Sharma, 2014, p.28).

Another feature of subcontracting in the livestock industry is that the products are much more standardised, a prerequisite for the commodification of the food animal. As noted, large meatpacking companies have a precise formula as to how their subcontractors should grow the animals. In a production contract, besides supplying the latter with the young animals, feed and veterinary support, the exact weight and fat content of each animal and how long they should be reared are tightly prescribed by the principal firms.

While contract farming in the meat industry has only gained global popularity in the past decade or two, it has long been recognised as a type of agricultural production predicated on 'contractual arrangements between farmers and companies, whether oral or written, specifying one or more conditions of production and/or marketing of an agricultural product' (Ewell, 1963, cited in Asano-Tamanoi, 1988, p.432). The debate over the merits of contract farming mirrors other forms of subcontracting (Neo, 2010). On the plus side, some argue that contract farming enables small-scale farmers to overcome their limited economy of scale (Heffernan, 1972). However, others argue that contract farmers essentially lose their independence and increase their exposure to risk (Althoff, 1979). Yet others have expressed concern about how socio-political control of local economic development is being eroded as a result of livestock subcontracting which sees activities being 'lifted out' and then recombined across larger units of time and space (Novek, 2003a, 2003b).

There is hence a risk when smaller farms become entrenched in the wider meat production networks and have little recourse to the kinds of (unreasonable) management demands and policies that are enacted from the top down. Many opponents of contract farming go beyond analysing economic gains and losses to look at the issue of social justice. As Watts (1994, pp.33–4) points out, 'the much-vaunted' independent contract farming functions as 'a little more than ... a hired hand on his or her own land'. Moreover, while contract farming is sometimes viewed 'as a savior of the family farm, and as a means to bolster agrarian populism', this view is highly debatable or, in the words of Watts, 'deeply ambiguous'.

In a critical review on contract farming, John Wilson (1986, pp.56–7) argues that 'contracting is one strategy whereby capital penetrates farming' and that contracts 'exploit the farmer and shift control of production out of farmers' hands into that of the processor'. Cutting across these sometimes divisive and polarised debates, one could conceive contract farming as a relational partnership between lead companies and contracted farms that is constantly evolving, with no necessary predetermined path – very much determined by the politics (of knowledge) surrounding such a mode of production. It is a perspective that recognises the potential for some contract farms to grow – after all, some of the biggest and most profitable farms in the United States are contractors to the big meat companies, for example, Cactus Feeders Inc. for beef and Iowa Select Farms for pigs, (Hendrickson and Heffernan, 2007) – and acknowledges the possible problems lead companies might face in such a mode of production. More importantly, Mariko Asano-Tamanoi (1988, p.433) argues that

> the way farmers perceive contract farming, i.e. defining, negotiating, and accepting 'contract' and 'contractual relationship' with food industries, differs in each cultural context, and that such perceptions can only be understood with reference to the economic, political and cultural forces to which they are inexorably linked.

In other words, for contract farmers to achieve some form of equity in their contracts, extra-economic concerns come to fore. It must be emphasised that the specific market strategies of contracting firms play a key role in shaping such perceptions too. Hence, among other things, there is a need to contextualise contract farming in particular places, with the understanding that, depending on the sociopolitical climate, different places would lead to different outcomes for the lead companies, the contracted farms as well as the local economy in general. For example, in a study of contract farming in the poultry sector in East Malaysia, Philip S. Morrison, Warwick E. Murray and

Dimbab Ngidang (2006) argue that given the structure of the poultry market, the cultural norms of farmers and the social-politics of East Malaysia, a state-administered contract farming programme is beneficial to the most marginalised people even as it probably would not cultivate long-term entrepreneurialism among them.

L. Labrianidis (1995), in his study of the Greek poultry subcontracting system, argues that while the extreme flexibility demanded by the principal processors is realised via subcontracting networks, it comes at the expense of numerous independent producers who essentially become cheap labour. More importantly, such flexibility 'does not lead to an upgrading of the work force'. The flexibility accrued is also the result of contract farms hiring cheap and 'essentially unskilled' workers or drawing underpaid help from the extended familial networks of the subcontractors themselves (Labrianidis, 1995, p.206). These implications point to the need to consider the fair remuneration of workers as well as their safety.

What we are perhaps witnessing in livestock subcontracting then is a return to an early merchant–employer system of subcontracting where the merchants have a high degree of control over their producers, with the latter often not in any position to negotiate better terms. As Pat Hudson (1983, p.125) notes, work on various processes of the production system is put out to 'an increasingly dependent, often landless, rural proletariat'. In other words, livestock sub-contracting is largely a form of cost-saving strategy that is a result of principal firms exploiting others to produce more than they could have produced on their own and, more importantly, at lower costs. Largely absent from the assessment of the contract system of meat production is the way it diminishes organic ties that farmers might have with food animals. The contract mode of production lends itself well to the commodification of food animals, such that they are seen as mere components towards the final goal. Contract farmers often do not have to track the growth of the food animals under their care and this contributes to a fundamental shift in the farmer–animal relationship.

Given the form of economic interaction implied in a contract farming system along with its contested costs and benefits, contracting firms need to establish direct local relations – itself a subtle form of governmentality – with a host of actors to convince the latter of the benefits of such a mode of production and its positive environmental and economic performance. Moreover, the nature of the industry (for example, the perennial concern about its impact on the local socioeconomic environment) also compels companies to engage the community. CSR statements and actions are thus imperative in the meat industry to, among other things, persuade potential partners that the firm is well meaning to the workers and to the community. To be sure, such statements and actions have to be well thought out and cannot exist at a purely discursive or vague 'public relations speak' level in order to achieve the ultimate goal of enrolling actors into their production system and logic.

Normalising Industrial Contract Farming: Case Study of Poland

FDI is seen as imperative in the restructuring and reinvigorating of moribund industries and for the development of new projects with export potential in developing nations. Poland along with Hungary and the Czech Republic are the top three countries for foreign investors in Central and Eastern Europe. By 1998 the total accumulated FDI stock in the 10 East Central European countries was $67,036 million, with Poland taking a 32.4 per cent share (Walkenhorst, 2004). Nonetheless, for Poland, investments are not evenly distributed, with regions near the eastern frontier receiving far less FDI than other parts of the country (Cieślik, 2005). For example, the north-western region, as a traditional pig-farming area, is a natural choice for foreign investors interested in this industry and has attracted almost all of the foreign pig investors. Yet the question is not just about how particular regions in Poland managed to attract FDI; it is also about how such regions can successfully transform themselves into more competitive and market-oriented entities and sustain such transformations. David Dornisch (2002), in his study of Poland's industrial region of Łódź, suggests that such a transition develops naturally through a variety of contingent projects, aided by local governments and institutions, and which eventually overcome its socialist-era inertia. Others too have pointed out the crucial difference local governments and institutions make in ensuring local economic development in the post-socialist restructuring and privatisation processes (Young and Kaczmarek, 2000). All these attest to the important role of public institutions in charting particular courses of development in the economic sector.

According to Eurostat (2010), Poland is Europe's third largest pig producer, after Denmark and Belgium. Since the beginning of the 1980s this figure has fluctuated a lot, reaching a high point of 22 million in 1992, but had shrunk by almost half to 11.4 million in 2012 (the lowest since 1964). While the number of pigs per 100 hectares of agricultural land in Poland (109 pigs in 2007) is higher than the EU average, it is significantly lower than that of the Netherlands, Denmark and Belgium. Over half of the pigs are held in the northern and north-western regions. For example, Wielkopolska is a hub that holds 30 per cent of Polish pigs. Essentially, some 5.2 million pigs are contained in a region with a sparse human population. Wielkopolska is among one of two Polish regions that experienced a substantial growth in the pig population, at 15 per cent per annum from 1999 to 2007. Moreover, small farms have diminished, having been replaced by large-scale farming. However, at the national scale there is as yet no dominance of large farms in terms of livestock. Only about 8 per cent of pigs in Poland are held in large farms of over a thousand pigs.

We have chosen the northern and north-western regions in Poland to enable a deeper understanding of how local governments, institutions and citizens respond to the investment strategies of two foreign meat companies (see Micek, Neo and Górecki, 2011 for a more in-depth analysis of this case study). Turostowo in Wielkopolska, western Poland, and Przechlewo in the north-west of the country were selected as they host two major foreign investors (the American-owned Agri Plus in Turustowo and Danish-owned Poldanor in Przechlewo) in the Polish pig industry and are located in two major adjacent Polish pig regions.

In one of its first acts as the government, the communist authorities (installed by the Soviet Union in 1944–5) tried to nationalise land. However, 'despite intense attempts at collectivisation of private farms into cooperatives in the period of 1949–56, land was kept mostly private in socialist Poland – an atypical hallmark of the country in the communist bloc' (Buchowski, 2003, p.47). In other words, under communism, agriculture was ironically the only key industry with a majority of assets privately owned. However, north-western Poland, our region of analysis, was distinct from other parts of the country because in 1990 half of arable land was state-owned (Dzun, 1991). Michał Buchowski (2003) argues that changes taking place in recent years are part of the chained history of Polish agriculture, and that what to scholars presents itself as a systemic transformation does not necessarily affect people's lives in a dramatic manner. We argue that this is true to a limited extent, because in some sites that we have seen the changes in agricultural practices affect people's lives substantively.

In the socialist era, large-scale (though not necessarily intensive) state-owned pig farms were constructed only from the 1970s in western and north-western Poland. They were said to be experimental farms as there was then no experience of such large-scale production. Under socialism, these farms were the largest sites of employment in the rural areas of the west and north-west. Such farms were essentially run by their respective directors who could be Polish United Workers' Party secretaries and were often local, well-known people. The emergence of a market-oriented economy has led to a shift in economic power to foreign hands, which can provide the necessary capital to expand the farms to compete in a more open economy. There are only a few Poles on the management boards, responsible mainly for public relations.

Socialist state-owned farms were rearing pigs in-house to the point of their slaughter and farms were not involved in the contract system. This is unsurprising given that it was rare then to have independent, entrepreneurial farmers. Where such farmers existed, they tended to conduct full production cycles (that is, rearing the pigs from birth to slaughter), and at a modest scale. With very few exceptions, state-owned farms were labour intensive and the machinery used constantly broke down. Vertical integration was the predominant mode of organisation under which state-owned farms sold pigs to state-owned slaughterhouses and then to public meat factories. Moreover, agricultural 'mini-towns', with their own infrastructure, were developed alongside these state-owned farms.

Under socialism, farms were viewed favourably largely because they employed many people. Not only were farms places of work for villages in the vicinity, they were often the focal point of local village life. For example, state-owned farms contributed to the social life of local people by supporting sports clubs and sponsoring practically all cultural events. In this way, the gatekeepers of the farms (usually their directors) tended to acquire social control over local people and ensure the governance of the social-economic order – an exemplification of the governmentality of social-economic lives. The farms thus had extra-economic value to the locale. Indeed, as will be discussed in the next section, foreign companies' attempts at building better community relations hark back to the socialist era.

However, by the early 1990s the majority of socialist state-owned farms were in financial difficulties because employees were not interested in reducing production costs or improving efficiency. They were also entrenched in inefficient and inflexible management systems that were unable to respond to changing market conditions (Woś, 1994). During the post-socialist era, both pig production and pig-processing factories progressively ceased to be state owned. Animex was established in 1951 as state-owned trading monopoly controlling all meat trade, both in and out of Poland. After the fall of communism in 1989, Animex decided to buy meat processing plants but was eventually bought over by the American-based Smithfield. As Morten Jensen, the Animex president, explains:

> Animex quickly became a big company that didn't have the know-how to run a meat business competitively … [and] was losing a substantial amount of money and required a company, with cash in hand, a certain philosophy and a long-term strategy to really be interested in taking over this company …
>
> (Deutsch, 2005)

Yet, as will be seen later, this is not just a matter of having the requisite technological know-how or capital. The successful governmentality of the post-socialist Polish pig industry by foreign interests involved other forms of knowledge transfer and political control.

In any case, Smithfield has effectively kept the world-renowned Animex name – best known in the United States for Krakus ham. Outwardly, the company appears to be the same, but it now operates according to Smithfield's management policies and focuses on profits. Nonetheless, even after the acquisition of Morliny (another Polish meat-processing company), sales were $600 million annually, merely a mid-sized company by European standards. By the mid-1990s Animex started to think about its own resource base because the meat from socialist-era meat factories had not often been of the highest quality. Animex proceeded to set up six finisher farms in Wielkopolska and north-east Poland. This production arm of Smithfield Poland was then named Agri Plus. Agri Plus entered the pig-production market by buying or leasing deteriorated state-owned farms from the socialist era.

In contrast, Poldanor, a Danish company, first entered the Polish market through leasing of deteriorated state-owned farms in 1994. Since then

Poldanor has continued to predominantly lease from the state. Animal production now takes place in 17 owned (or leased) farms and five contract farms are run in cooperation with local farmers. In 2010 Poldanor employed 540 Poles and six Danes. In terms of specific modes of production, both Agri Plus and Poldanor adopt a three-site production concept, in contrast to the method of production in the socialist era. Essentially, the production cycle is divided into three stages: sow and piglet breeding farms, weaner farms and finisher farms. Both firms have all their operations concentrated in northern and north-western Poland. The key difference between Agri Plus and Poldanor is that the latter is less keen on contract farming, preferring to be in full direct control of the production process. In contrast, contract farming – under the auspices of its contract growers' scheme – remains a cornerstone of Agri Plus's strategy for growth.

Tensions and Perceptions in Polish Pig Production

While such a fundamental strategic difference reflects the different cultural business models of the respective lead firms, the context and history of the specific region influences the receptivity of local communities to both the firms' strategy and presence as well. In this case, the overall reactions to the companies are dissimilar in that the communities surrounding Poldanor's operations believe that the pig farms have a significant positive impact on the local economy – in terms of employment. This contrasts with Agri Plus. Indeed, in an interview, an Agri Plus manager admitted that pig farms do not create much new employment compared to other industries. In the case of the farms in Turostowo (see Figures 2.1 and 2.2), the only obvious impact has been contracts with a few individual farmers to harvest pig manure on their arable lands. However, revenue from this cooperation constitutes a small share of their income. This cooperation is thus perceived by local people as unprofitable.

Clearly, automated finisher farms do not require large employment: it is not uncommon for one worker to tend to more than 1,000 finishers due to automatic feeding and ventilation systems. In former state-owned finisher farms, one worker looked after fewer than 200 pigs. Hence the presence of large-scale intensive farms does not solve the problem of rural unemployment. However, there is one exception, in Przechlewo commune, where spatially concentrated production chains have generated comparatively high employment; the unemployment level is comparatively low, at 10.7 per cent in 2007. Poldanor and Prime Food (a meat-processing factory controlled by the same Danish investors in Poldanor) together employ about 47 per cent of the total number of employees in the community.

As of 2009 there were 39 large foreign farms in Poland. Farms managed by these two foreign enterprises are located mainly in Zachodniopomorskie and Pomorskie provinces in the north-west (see Figure 2.3). In 2007 the lowest ratio of pigs held by individual farmers was reported for these two regions

Figure 2.1 Typical pig farm in Turostowo, Poland

Figure 2.2 Typical storage facility for animal feed in Turostowo, Poland

(66 per cent and 75 per cent in comparison with an average of 87 per cent), suggesting that large corporatised farms are making headway in this region. Less than one-third of Agri Plus and Poldanor farms are situated in four other provinces (Wielkopolska, Warmińsko-mazurskie, Kujawsko-Pomorskie and Mazowieckie). Agri Plus operations are relatively geographically dispersed in Zachodniopomorskie (nine farms), Wielkopolska (five farms), Warmińsko-mazurskie (four farms) and Kujawsko-Pomorskie (one farm). Poldanor's farms are spatially concentrated; the vast majority of them are located in Pomorskie (mainly in Człuchowski county) and Zachodniopomorskie province. Only two small new pig farms (in Karsy and Mokrzk) are situated in Mazowieckie, central Poland.

A key tension between global and local interests is seen by the completely different perceptions of the role of the contract grower programme in the development of the Polish pig industry. Agri Plus argues that one key growth strategy to propel the local industry is through its contract growers' programme. Started in 2002, Agri Plus states that it chooses farmers to join the contract growers' scheme based on three criteria: the willingness of the farmer to run the farm; the farmer's experience and knowledge; and the health and safety conditions of the farm. After a relatively short period of seven years, there are now almost 400 farmers involved in the Agri Plus programme. The company's goal is to pursue a model contracting system that showcases

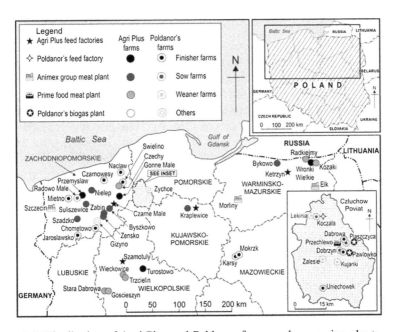

Figure 2.3 Distribution of Agri Plus and Poldanor farms and processing plants
Note: Only Agri Plus or Poldanor-owned facilities are shown. Contract farmers of the companies are not shown.

modern agriculture in Poland. The contract farming system, in other words, is sold and marketed as a necessary organisational innovation in today's competitive meat market. On its website, Polish farmers are quoted, praising the scheme and emphasising the key positives of the scheme as 'stability', 'reliability' and having the scope for expansion (Box 2.1).

Box 2.1 Farmers' perceptions of the contract grower programme

'We joined the contract grower program in 2003. We are satisfied with the cooperation because it helped us to stabilize our financial situation. Now we are thinking of increasing the number of contracted finishing places.'

'I started on a small scale two years ago. It was worth doing. Apart from a regular payment per hog I got an extra bonus for a low feed conversion rate. So I decided to expand my production on a second farm. The company is reliable and I can always count on their support.'

'What I appreciate most in the cooperation with Agri Plus is stable cash flow and no market risk. I get my monthly contract payment regularly paid into my bank account. I can always make a profit regardless of a current market price for hogs.'

'I learnt about the program from a farmer who signed the grower contract earlier. Cooperating with Agri Plus has helped me to expand my animal production and create new jobs. Receiving a predictable income has encouraged me to start thinking about new investment and the purchase of modern equipment.'

Source: Agri Plus Poland (http://www.agriplus.pl/)

Agri Plus has taken care to emphasise guaranteed sales stability and independence from the risk of price drops to the farmers. Moreover, there are additional bonuses for the high quality of meat provided. Agri Plus also supplies feeders and feeds for finishing, veterinary support and ensures production control and the transport of animals. It is through the promulgation of such knowledges, discursively constructed publicity and institutional support that contract farming as a new form of governmentality of the pig industry took root in Poland.

Nonetheless, there are some farmers, especially those not involved in the programme, who are highly critical of the scheme. While acknowledging the possibility of income stability, they nonetheless feel that specific regulations in the contract are not favourable to them. First, it is not easy for the farmers to terminate the contract. Second, they are wary that the proposed length of contracts can run as long as 12 years. In fact, several farmers we interviewed remarked that their lawyers have all advised against signing the contracts. Third, some elderly farmers are too used to existing ('traditional') mode of production; that is, from the animal's birth to its slaughter.

For its part, faced with such initial scepticism and reservations from some farmers, Agri Plus has significantly modified production chains in order to attract local farmers to enrol. For example, they have chosen to predominantly contract out the finishing stage (in other words, contract farmers are tasked only to fatten the pigs). However, there are still farms where foreign companies have not been able to introduce any changes due to local resistance. In Turostowo, all local farmers in the 5-kilometre area surrounding the Agri Plus facility refused to become subcontractors. Yet the steady growth of the contract farming scheme is undeniable. While Agri Plus used to supply only 10 per cent of Animex's demand for pork in the early 2000s, by 2010 it was able to supply almost 40 per cent, with little expansion of its own farms. This indicates that Agri Plus is able to obtain more pigs from its contract farmers.

For Poldanor, its main activity is pig breeding, including production of farm gilts (young female pigs that have yet to farrow) under a contract with Pig Improvement Company for sale in Poland and abroad. The production of pigs in Poldanor is based on breeding material imported from Britain and Denmark. The firm does not rely on a contract farming model and, in contrast to Agri Plus, concentrates its operations in a few select places.

As mentioned earlier, the emergence of capitalist relations has shifted power from local to foreign hands. While there are Polish directors of farms (in charge of public relations), they often come from outside the local neigh-bourhood. This is a significant departure from the socialist-era farms where local people managed the farms. As far as the local authorities are concerned, global players tend to carry out an isolationist hands-off policy towards them. Incidentally, this marks a possible point of difference in attitudes towards Poldanor (a small European-based company) and Agri Plus (owned by the then biggest and most profitable pig processor in the world, Smithfield). According to the authorities interviewed, companies do not inform them about important events on the farm and limit contacts with the authorities. Ironically, likely strategies to control and manage foreign companies might lead to local authorities further alienating the former. These initiatives include the introduction of limits to the number of pigs per hectare, local spatial zoning and a ban on non-bedding production. Relations between farms and local communities are shaped by actions taken by foreign companies under their CSR programmes. They constitute the local social and institutional embeddedness of global companies. The official aim of global players is to realise the rules of good neighbourliness and sustain positive relations within the local community.

The reality of social actions of global companies is more complex. In some places (Turostowo) the support of foreign companies is somewhat favourable; in other places it is almost antagonistic. However, Poldanor attempts to cultivate better relations not only by helping local schools and sports clubs but also spear-heading local environmentally friendly initiatives. For instance, Poldanor con-sistently supports the cleaning of local rivers. Furthermore, the building of biogas plants (four by 2010) is well regarded by the local community as an act of good corporate citizenship. To be fair, the geographical concentration of Poldanor's operations lends itself better to the cultivation and maintenance of good community relations, as compared with the more dispersed operations of Agri Plus. Indeed, Poldanor's CSR actions are reminiscent of the socialist-era farms which were the focal point of community life. While Agri Plus is moving similarly in the right direction in terms of CSR, the fact that it has a larger number of authorities to cooperate with across various communities in different locales could give rise to the perception that their actions are scattered and ad hoc.

In sum, both Poldanor and Agri Plus have operations in Poland which are markedly different from those of the socialist period. Their 'advances' may be viewed as necessary given the perceived less-than-efficient modes of production in socialist times. Indeed, as many have pointed out, foreign investors not only bring fresh financial flows but also new methods of production, new technologies, and improvements in organisation and management techniques (Dadak, 2004); in other words, a new form of governmentality. The latter extends to the social milieu of the farmers as well as through specific CSR initiatives. Nonetheless, due to the differing corporate philosophies of their parent companies as well as different opportunities that existed at their point of entry into the pig industry, Poldanor and Agri Plus differ quite fundamentally on their expansion strategies and production mechanisms. Yet, for both companies, CSR programmes are all the more critical given the historic and central socioeconomic roles played by state-owned farms in their invested regions.

While resistance to the entry of foreign companies into the community often fades after the initial years (because of the lack of recourse and resource), building better relations, along with convincing farmers and residents that the 'modern' way of producing meat is more beneficial for everyone, remains an ongoing process. Not least, it is important to minimise future protests and political opposition should relevant legislation come into force that might restrict the companies' developmental strategies. In this regard, Poldanor seems to be doing better in immersing itself into the institutional environment of the locale. While no doubt aided by its strategic decision to concentrate its operations in only a few locations, Poldanor appears to appreciate more the extra-economic roles of pig farms in rural areas. On the other hand, Agri Plus not only has to contend with a similar aim of building ties with the community, it has the added task of persuading more farmers of the wisdom of a contract mode of production. In that sense, Agri Plus is more explicitly spearheading the restructuring of the Polish pig industry along lines similar to that which exists in North America.

Whatever the case, it is clear that despite some nuances the Polish authorities on the whole welcome such investments which ultimately transform and normalise the pig industry to one that is highly intensified, commodified and efficient.

When Culture Meets Political Economy: The Contrary Case Study of Malaysia

The experience of Poland is not representative of all places. As alluded to earlier, the transformation of the techniques of production is purposeful, and resistance to such changes must be managed and ameliorated. In other places, the homogenisation trend of livestock industry restructuring is unable to take root in the first instance due to cultural politics. This is illustrated through the example of Malaysia, a multireligious country located in Southeast Asia with a Muslim majority (61 per cent) and a sizeable Chinese minority (24 per cent).

Malaysia is governed by a coalition led by an exclusively Malay political party, United Malays National Organisation (UMNO). Despite widely acknowledged institutional discrimination against the Chinese (and other minorities), until fairly recently overt displays of racism were uncommon, as were outright accusations of racism (Gomez and Saravanamuttu, 2012). Indeed, all major political parties in Malaysia claim they value the imperative of religious and ethnic harmony. UMNO, together with its coalition partners of other ethnically based parties, has governed Malaysia since independence in 1957, and has survived setbacks in national election polls in 2008 and 2013. The main opposition coalition, organised first through Pakatan Rakyat (People's Alliance) and then Pakatan Harapan (Hope Alliance) since 2015, is even more explicit in its ambitious multiethnic approach towards governance. In other words, at the broadest *rhetorical* level, politics does not deviate from a multiracial model of governance and ethnic harmony is deemed paramount at both ends of the political divide. Nonetheless, even such rhetoric has been severely compromised in recent years, with both UMNO and the opposition Parti Islam Se-Malaysia (PAS) competing to capture more of the Malay-Muslim vote through ratcheting up their respective ethnoreligious credentials. There are even calls from both parties to enact strict *shariah* laws, as are already in place in the east coast state of Kelantan (Moftah, 2015). In response to these developments, political commentators argue that the secular Federal Constitution has been contravened as well as the fundamental freedoms of all Malaysians in a sharp turn towards authoritarianism (Azrul, 2015). In any case, at the local, vernacular level, subtle (and not-so-subtle) racism and the continual assertion of a cultural and religious boundary that separates the Malay-Muslim majority and the Chinese minority can be discerned and is arguably deepening. Such underlying racial politics significantly affects the development of the Malaysian pig industry.

The following discussion on the governmentality of pig farming in Malaysia was based on field research conducted in Malaysia for a total of nine months in 2005, 2006 and 2009 and is drawn in part from Neo (2009) and Neo (2012). We show how Chinese farmers and pig farming are racialised through governmentality by highlighting two related issues: the siting of farms and regulatory control over the function of the farms. Concomitantly, we show how these issues produce (and are reproduced by) racial and cultural discourse and how they ultimately change the course of the development of the pig industry in Malaysia.

As Nikolas Rose (2006) explains, governmentality is an iterative process between (official) regulation and subjects' ways of life that is undergirded by governing technologies, knowledges, culture, as well as differing power politics. The question of the siting of pig farms is, for example, tied to the question of farmers' land rights which in turn involves a different govern-mentality of its own. Such governance is marked by multiple and inchoate scales of control and vagueness in both the spirit and the letter of the law, making the siting of pig farms fraught with difficulties. The following discussion develops this argument.

In Malaysia, agricultural affairs are devolved to state-level authorities. Hence, for pig farms the governing body is the state-level Department of Veterinary Services (DVS) and not its federal equivalent. Yet the state DVS is actually only concerned with the environmental performance of pig farms. Indeed, the environmental pollution problems that specific farms might have had were never severe enough for the DVS to force their closures (see Figures 2.4 and 2.5). Rather, the primary reason/justification for shutting down pig farms has always been that the land they were sitting on was needed for alternative development; or that the land was not used according to its proper designation. Such demands are issued by the Development Office and/ or the Land Office.

At the federal level, there have been frequent calls to the state governments to centralise pig farming in each state (a likely prerequisite for the further intensification of pig farming), with the most recent call for this occurring in 2006. However, by 2007 it was officially announced that the state governments had once again stayed the decision to make all pig farmers rear their livestock in centralised pig-farming areas. This was because many state governments had difficulty identifying suitable land which was far from residential areas but easily accessible at the same time. Indeed, any proposed relocation site was vehemently opposed by those already living in the vicinity. For example,

Figure 2.4 Nursery pigs at a pig farm, Malaysia

Figure 2.5 Weaning piglets at a pig farm, Malaysia. Note the restricted space for the sow

in 2008, when a site was but merely mentioned in passing as the possible place for the relocation of pig farms in the state of Selangor, an anonymous Malay reader wrote to the newspaper *Utusan Malaysia* to protest:

> [Also], why are they carrying out the pig farming project in the Sepang district when the majority of its residents are Malays? Muslims are not the ones who are going to consume pork, so why didn't they choose an area that has a majority of Chinese residents? While the Chinese are savouring pork, the Malays are forced to smell the odour and are faced with a pollution problem.

The writer draws on the culturally rooted gastronomic pleasures of the Chinese and juxtaposes them with the 'sufferings' of the Malays. Such boundary-making is a false dichotomy as it presumes that others (including non-Malays) who live close to pig farms do not suffer (or do not mind) the same odours and pollution. Nonetheless, the reader's view illustrates that having a permanent and secure place to farm is fraught with difficulties even though it is critical to the sustainable future of the pig farmers. Among other requirements, what is at stake here is the provision of secure land rights to the Chinese farmers. However, the issue of land rights, in this case, is embedded in the broader ideology of *bumiputera*. *Bumiputera* literally means 'sons of the soil' and is a cultural construct that is used to legitimise the special status of native Malays (and others) as the 'indigenous' occupiers of Malaysia. This entrenched notion (which is the basis of the affirmative action in favour of the Malays), in an Orientalist fashion fuels the idea, even among the Chinese, that 'the

Malays are the host and what benefit we Chinese get, we get on their terms, not ours' (Ye, 2003, p.55). In this way, land rights become a cultural-political question, as much as it is a legal and development question.

The entwining of culture with the pig industry rears its head in other ways. Malay politicians are wont to play up the cultural roots of pig rearing to paint a picture of moral panic where pigs are produced in such large numbers as to defile the land of the *bumiputera*. For example, a state assemblyman from Selangor remarked in a newspaper interview headlined 'Total pigs reared exceed locals':

> 'In this area, there are already more than 9,000 Chinese and every one of them usually rears 25 to 30 pigs. Imagine if every pig is able to produce 10 piglets in three months, making it 40 pigs in one year,' said Datuk Dr Karim Mansor, State Assemblyman for Tanjung Sepat.
>
> (*Berita Harian*, 2008)

Given these recent developments, it may come as a surprise then that the early years of commercialised pig farming were relatively free of cultural and political inflections. Indeed, there is evidence that the government viewed pig farming as just another economically productive activity (treating pigs essentially as a commodity) that helped in the development of rural areas and raised the standard of living of poorer Chinese especially. Farmers whom we have interviewed noted that only in the late 1950s did pig farming progress from a backyard activity to small-scale commercial farming. All remembered the days when state governments actively helped grow the industry. In a 1965 report on the development of the pig industry, the Ministry of Agriculture (1965, p.78) described the pig scheme (initiated under the auspices of a rural development livestock scheme) in the following terms:

> Groups of 5–10 people are selected; they are required to put up a shed according to a specific plan recommended by the department, with financial assistance to the extent of $50 worth of building material to each farmer. Once the sheds are completed, each farmer is given 5 good quality weaned piglets.... The farmer is permitted to sell 4 of the 5 pigs at marketing age and retain one pig as a breeding sow. Out of this one sow, he should return to the government 5 gilts [that is, young sows]. These 5 gilts in turn are distributed to a new group of farmers.

Several years later, in 1971, the federal DVS even organised a symposium on the rearing of pigs. In his foreword on the published proceedings of this symposium, the director-general of the DVS noted that 'the pig industry constitutes a very important component of animal production in Malaysia' and 'firmly believed' that the symposium had 'gone a long way not only towards a greater understanding of the industry, but also in providing a realistic awareness of the problems that require remedy' (Devendra, et al., 1972). It is telling that the chief problem identified was the fluctuating prices of pork. The symposium recommended that, among other things, central and regional abattoirs needed to be established so that slaughtered pigs could be more 'marketable' overseas through higher levels of hygiene and food safety. Indeed, the report, co-written by a Chinese and a Malay, goes on to note that unless 'proper marketing' were undertaken 'the expansion of this multimillion dollar industry would be retarded' (Hussein and Lim, 1972, p.141). Furthermore, under the ad hoc system then operating it was thought that middlemen acting for slaughterhouses were benefiting at the expense of farmers and consumers who 'always get the worse share of the bargain' (Hussein and Lim, 1972, p.145–6). These views all suggest that the Malaysian pig industry *was* at the cusp of being drawn into a more intensified and technologised mode of production. In other words, in the years when pig farming shifted from a backyard to commercial mode of production, there was a relative absence of any popular (cultural or otherwise) opinion about it and all signs pointed to a commodified perspective towards pigs and pig farming. Attitudes towards the industry glimpsed through official reports were thus largely non-ideological (and arguably positive), with the pig industry being treated as a *bona fide* rural economic activity.

By the 1970s, however, the largely unregulated growth and increasing lack of 'appropriate' space for farming pigs began to present problems for government officials. A 1975 report noted that 'the lack of suitable land for pig keeping is becoming an increasingly acute problem … causing socio-religious problems' (Ministry of Agriculture and Rural Development, 1975, p.56). In a separate report published by the United Nations Development Programme a year later, it was similarly stated that the proliferation of pig farms was causing 'religious and social problems' (United Nations Development Programme, 1976, p.59). Neither report elaborated on the specifics of such social and religious problems. However, we can safely assume that these problems largely referred to growing Muslim sensitivities towards pigs, and the increasing difficulty in situating farms away from residential (especially Muslim) areas.

As testament to this nascent objection, a long-time senior federal veterinary officer remarks: 'You know what my previous boss [that is, the head of the Pig Unit] used to say back in the 1970s? He said that the pig industry is not to be seen, not to be smelled and not to be heard. The last thing we want is problems from the industry' (interview, 12 January 2006). This admission suggests that it was probably in the 1970s that the pig farms emerged to become an unspoken, concretised site that metaphorically divided the two largest ethnic groups in Malaysia. One particular farmer remembers the 1970s as such:

I actually don't know when or how it happened, but it came to a point then [in the 1970s] that I started to feel that I am doing something wrong by rearing pigs. It is not that there were protests or anything like that, but that feeling, for the first time, of being different and discriminated simply because of what we are doing.

(Interview, 10 June 2006)

Similar to the question of land rights, other forms of regulation and control are often outcomes of a social-cultural and racialisation process. From the mid-1970s to the early 1980s, a period that mirrored the global rise in Islamic consciousness, four Malaysian states passed enactments (equivalent to state laws) to oversee the pig industry within each state. These states were Johor (enacted in 1975), Terengganu (enacted in 1976), Negeri Sembilan and Melaka (both enacted in 1980). While ostensibly meant to manage the pig industry, these laws in effect further concretised the division between the Chinese and Malays. In other words, the pig industry became an economic activity that is different from others and one which is to be largely identified by its affinity to one ethnic group and by its repugnance to another (religious) group.

The irony of this movement from a commodified view of pig farming to an activity that is culturally and religiously inflected is that it can also become a tool for the marginalisation of pig farmers. In other words, seeing pigs and pig farming as more than commodities actually gives rise to other intractable problems. For example, in subsection 2 of section 4 of the Melaka Pig Enactment Act, no pig farms 'shall be erected within one hundred and sixty metres of the precincts of a Muslim dwelling house and five hundred metres from any building used for the purpose of worship by Muslims'. Subsection 1 of section 6 notes: 'If any pig is at any time found straying into the compound or premises of any dwelling house occupied by a Muslim, or used for the purpose of worship by a Muslim, its owner shall be guilty of an offence against this Enactment'.

Viewed plainly, such 'zoning' is arbitrary; for example, why 160 metres for Muslim households and 500 metres for places of worship? But more importantly and practically, this zoning greatly limits the sites in which farms can be (re)located. In the late 1999s the Malaysian pig industry suffered a particularly severe virus attack and it was in this trying period that cultural constructions and discourses over pigs and the pig industry became more explicit than ever. By May 1999, just nine months after its mysterious appearance, the Nipah virus (Nipah encephalitis) had all but crippled the industry. The outbreak resulted in the deaths of 125 pig farm workers and the culling of 1.2 million pigs (representing half of the total pig population). An estimated 36,000 people lost their employment and total economic costs were estimated to be $520 million. Not only were the impacts of the Nipah virus extensive, it also brought the pig industry back into the limelight. For the pig farmers of the state of Negeri Sembilan (which was then the biggest pig-producing state), the impact of the Nipah virus was fatal. Every single pig in the state was culled and pig farming was subsequently outlawed in 1999. By the end of the crisis, the total number of pig farms in Malaysia was reduced from over 3,400 in 1997 to 672 in 2005.

Although it is believed that fruit bats were the secondary vector of trans-mission, 'the early epidemiology of the disease in Perak and the spillover mechanism that first introduced the infection to pigs remains undetermined' (Food and Agriculture Organization, 2002, p.4). Interestingly, some among the pig farming community are adamant that the Nipah virus is a mutation of the Hendra virus (HeV) which more commonly afflicts horses. In what is held to be an ironic twist of fate, they speculate that the source of HeV was from the importation of Australian pedigree horses by the royal house of Perak state. As the Malay royalty was involved, these horses are said to have never been subject to the usual quarantine process that might have revealed their infected status. A farmer wryly asserted: 'They say the pig got the disease because it is unclean, but the truth is the virus is from horses which were secretly smuggled by the Malay royalty. The Malays literally killed the pig industry!' (interview, April 2006).

However, the dominant discourse in the aftermath of this catastrophe is to pathologise the entire Chinese community. While this does not go so far as to 'read disease and depravity into ... the political anatomy of Chinese', as has been argued in the case of the smallpox epidemic in San Francisco in the late 1800s (Craddock, 1999), there were nonetheless strong views being expressed in the Malay press. A Malay reader wrote to *Utusan Malaysia* lambasting the Chinese farmers and evoking the metaphor of purity, thereby reinforcing the divide between Chinese and Malay-Muslims through beastly and cultural racialisation (Zaid, 2000):

> If the [Nipah] virus spreads, they [pig farmers] should take full blame because they do not have the expertise to rear that forbidden animal. Sometimes, those Chinese who eat pork do not understand how disgusted the Muslims are towards pigs. They only know that Muslims do not eat it and it is forbidden. But they do not know what is felt by the Muslims, especially the Malays.

Yet the notion that pigs are a sensitive issue is not merely psychological or ideological. Such sensitivities impact on the way pig farms could be run and developed in Malaysia, particularly since the 1980s. For example, it was about then that manure collected could no longer be used to fertilise other agricultural lands as evinced by the words of Ahmad Mustaffa Babjee (1983, p.iv), chairman of the ad hoc pig waste pollution technical committee set up in the early 1980s: 'In other countries where Muslim sensitivities are absent, the disposal of sludge is relatively easy as it is a good source of fertiliser. The pollution control of pig wastes in Malaysia thus poses a unique problem which requires in-depth study'.

The experience of pig industry development in Malaysia, with its strong social-cultural underpinnings, is arguably an exception to the rule. In most other places, objections and controversies surrounding the siting of livestock farms are more often rooted political-economic (as seen in Poland) and environmental concerns. Nonetheless, it presents a clear contrary case of stagnating livestock development due to local cultural politics and concomitantly the shifting policy position of the government. In this case, there is no complicity between the government and the industry.

Conclusion

This chapter argues that while the global trend in the livestock industry is towards heightened intensification, by employing organisational innovation like subcontracting, and lowering costs, by exploiting low-cost production sites, such a spread of the political economy of the livestock industry does not proceed unfettered. In most cases, the complicity of national and regional governments is required. Even then, the local community still needs to be convinced of the positive effects of such a change to the way they rear food animals because the power of the governing institutions is not total. The Polish pig industry is used to illustrate the ways in which global companies convince the local community to enrol in a particular mode of production, invariably towards greater intensification and modernisation. However, the assumption that food animal production will continue to be intensified does not hold in all places. This is especially true when we consider issues beyond the economic and the regulatory. With rising religiosity and the willingness of the governing institutions to play the politics of religion and culture, places like Malaysia have seen a decline in some livestock production. Ironically, the unintended effect of this is the denormalisation of factory farming in the pig industry. Beyond this specific case, it also suggests the critical role governance and regulation play in shaping the development of the livestock industry and also the conceptual scope to discursively construct food animals (Neo, 2010).

In the next chapter, we discuss how other institutions draw on technology to commodify food animals and normalise factory farming, albeit to a varying extent, with the support (tacit or otherwise) of regulatory structures described in this chapter.

3 Science, Technology and the Commodification of Food Animals

Introduction

Following from Foucault's sense of biopolitics as the intersection of life and politics, we see the direct correspondence between the commodification of animal bodies and food politics, at least in Europe and North America. This particular biopolitics involved the production of multiple disciplines, 'truths' and subjectivities – those of the animals, the farmers, the scientists and the consumers. Since at least the Second World War, the science and technology of livestock production has focused upon producing more flesh, more eggs, more milk and more offspring as efficiently as possible. The unilateral focus of livestock science on productivity is awe-inspiring through its myopia. Reading through years of research output via scholarly journals one finds almost no mention of keeping farmers on the land or keeping workers in their jobs. Only recently have environmental and animal welfare concerns been voiced (see Chapters 4 and 5), and those only because of the social movements bringing them into the public arena. The single-mindedness of the group of disciplines that make up the animal research enterprise has shaped the pathway leading us to this place where livestock production is problematic in nearly every aspect except for productivity. Even as recently as 2013 a leading livestock scientist wrote that 'technology-induced food price declines unambiguously benefit consumers' (Lusk, 2013, p.22), ignoring all the public health warnings about too many livestock products in the diet or the large numbers of environmental issues emerging from livestock production. The 'benefit' alluded to is about dollar price alone and is blind to an array of negative externalities that are concomitant to this 'benefit'.

The current animal science research literature is awash with concerns about providing food for the 7.5 billion people on the planet. Many papers published in *Animal Frontiers*, a review journal of the American livestock industry, begin with a cautionary nod to the nutrition problems of the poor and the challenge of feeding the world in the future. But this is as far as the reasoning for increased production or productivity or livestock products goes. That is the orientation – even the single orientation – driving the new age of genomics. The international policy literature is not so positive about the upwardly trending numbers globally. The World Bank, United Nations Food and Agriculture Organization (FAO), International Livestock Research Institute and other institutions are cautionary about the expected environmental and public health issues associated with a doubling of livestock production over the ensuing decades, although they are still proponents of livestock as a route to 'development' (Steinfeld, et al., 2006; World Bank, 2009; Thornton, 2010). In particular, the waste issues and zoonoses resulting from concentrated animal factories are of grave concern. Various publics and activist groups are promoting significant changes in animal welfare conditions.

Agriculture colleges, at least in the United States, increasingly depend on the pharmaceutical and livestock supply corporations for research grants, a sizeable portion of which cover overhead and administrative costs. More than two-thirds of animal scientists reported in a 2005 survey that they had received money from the industry in the previous five years (Petersen, 2012). Indeed, one of the major problems with nearly all livestock animal science journals is that authors are not required to identify their research sponsors on their published papers. Private research funding increasingly outweighs public funding and that is probably even more the case in countries with insubstantial tax bases. The private sector has provided most of the inputs and knowledge for growth in the confined poultry sector, at least in Africa, where breeding stock, medicine and feed are the most important factors in explaining broiler productivity growth (Pray, Gisselquist and Nagarajan, 2011; Waithanji, 2015).

Animal science has largely developed, especially in the United States, as a driver of national development but, more importantly, as a manufacturer of greater profits, particularly for the owners of large farms, pharmaceutical and feed companies, breeders, processors and retailers. Much of the research performed in state agricultural colleges, funded by taxpayers as well as meat-producing and attendant or supply chain corporations, contributes to large producers, processors and corporate entities. The animal–science–industry complex is hence in part an outcome of the complicity of private–public interests.

The outline of this chapter includes some introduction to concepts that help understand the technoscience that is an important driver of the industrial livestock sector. Our basic framework includes the animal body through the lens of Foucault, the animal body as commodity using Marxist theory, and a political economy of science after Jack R. Kloppenburg and other sociologists and historians of agricultural science. These are all refinements, elaborations and applications of the key concepts introduced in Chapter 1. Following these introductory sections, and a brief overview of the history of animal science, we consider the ways in which the commodified animal body is pushed to yield faster and more, with the aid of new technologies. We also consider research on killing, 'depopulation' and rendering (the latter being the process that turns food animal wastes or by-products into other uses).

In the field of animal genetics, a select group of companies dominate the playing field. Bull semen, artificial insemination and genetic markers (single nucleotide polymorphisms, SNPs) for genetic disease and meat quality have been available for some time. Boar semen, which breaks down when frozen, has not been widely available but some advances are being made in research and development of semen life extension for the market. According to Keith Fuglie, et al. (2011), the poultry breeding industry has undergone extensive concentration over the past two decades. Broiler breeding is dominated by three firms and both layer breeding and turkey breeding are dominated by two. The EW Group, Groupe Grimaud and Hendrix Genetics are all European-based, multispecies genetics companies. Cobb-Vantress, a subsidiary of Tyson Foods, is a broiler breeding specialist. Cobb-Vantress, Aviagen Broilers (EW Group) and Hubbard (Groupe Grimaud) together supply at least 95 per cent of the global commercial breeding stock for broilers. Perdue's Heritage Farms sells broiler breeding stock only in the United States, most recently to Cobb-Vantress. The EW Group and Hendrix Genetics supply almost all of the global breeding stock for layers and turkeys. Regional breeding firms (in India and elsewhere) tend to work through joint licensing agreements with one of the bigger firms. Concentration has resulted from higher fixed costs due to molecular biology advances and the economies of scale that can be captured from size and geographic scope. Multispecies capacity also provides for economies of scale.

The pig genetics industry is less concentrated than the poultry industry but has still seen consolidation over the past decade with the further development of molecular biology and genetic science. The leading global pig genetics company is PIC, owned by Genus, a publicly traded British firm that also invests in cattle breeding through its US-based subsidiary, ABS Global. Smithfield Premium Genetics, a subsidiary of the vertically integrated pork producer and processor Smithfield Foods, supplies pig breeding stock internally to producers in the Smithfield system. Hypor (a subsidiary of Hendrix Genetics) and Newsham (a subsidiary of Groupe Grimaud) also have significant pig breeding research and development (R&D) and international sales. Other important pig genetics suppliers include two farmer-owned cooperatives, DanBred (Denmark based) and Topigs (Netherlands based), which also export breeding stock to other countries. Nonetheless, farmer cooperatives and national breed societies still maintain an important place in providing breeding stock unlike the poultry industry (as can be seen in Chapter 6 when we discuss the example of organic pig farming in south-west China).

Beef and dairy cattle research and development centre on artificial insemination, embryo transfer, sexed semen, genetic marker technology and SNP chips developed through a collaboration involving a genetics sequencing company whose primary focus is human health (Illumina), the US Department of Agriculture (USDA), American and Canadian universities, and the major North American artificial insemination companies (Fuglie, et al., 2011). The artificial insemination industry, at least in the United States, has undergone similar consolidation as the pig and poultry industries, with only five companies providing 90 per cent of the processed bull semen in the early 2000s. The embryo transfer business in Latin America and Asia has grown rapidly over the past decade. Private research spending on genetics across all species was estimated at approximately $300 million for 2006–7, with nearly half spent on poultry (Fuglie, et al., 2011). Nearly 98 per cent of the R&D was sponsored by North American and European companies and cooperatives.

For such leading meat conglomerates, the search for bigger markets to exploit has extended beyond national boundaries in the past decade. The Brazilian meatpacking company JBS has since 2007 bought out several leading meatpacking companies in the United States, including the beef business arm of Smithfield Foods. As noted earlier, the rest of Smithfield Foods was purchased by WH Group (previously known as Shuanghui International Holdings), the biggest China-based meatpacking company. Other post-socialist countries in Central and Eastern Europe which are transitioning to a more market-based economy also offer immense possibilities for investment. Smithfield has, in the past 20 years, made several investments in Eastern Europe, as have Western European meat companies like Danish Crown AmbA (see Chapter 1). The immediate implication of such transnational investment from developed countries to less-developed countries is the likely transfer of new technologies and knowledges of producing meat. Such a transfer will have significant consequences on the structures of the meat industry in these less-developed countries. For instance, and as has been discussed, mirroring the trend in the developed countries, contract farming as an organisational feature of meat production may become more prevalent. One key, yet difficult, question regarding such investments is in what ways do they benefit (or not benefit) the various actors in the industry. An equally important question is how contract farming as a form of production impacts on non-human actors. Indeed, one can see contract farming as a form of governmentality which fundamentally changes not only the practices and policies of farmers and planners by convert politics but also the spread of specific scientific knowledges and technologies that transform the animal body.

Biopower, Biopolitics and Chained Commodities

Animal bodies are managed in a variety of ways with the goal of maximising profit. Managing the bodies of the animals to produce more flesh, milk and eggs in the most efficient way is the basic goal of industrial livestock production networks. Following other livestock science analysts like Carol Morris and Lewis Holloway (2009) and Richard Twine (2010), we find the lens of biopower and biopolitics an apt framing of the current (and past) technoscience efforts to achieve this goal. Foucault wrote about the development of two poles of biopower in the seventeenth century. The first pole to develop

> centred on the body as a machine: its disciplining, the optimisation of its capabilities, the extortion of its forces, the parallel increase of its usefulness and its docility, its integration into systems of efficient and economic

controls, all this was ensured by the procedures of power that characterised the *disciplines*: an *anatomo-politics of the human body* ...

(Foucault, 1990, p.139)

The second pole has to do with what Foucault calls the 'species body' or the body of the population – its birth, death, propagation, health, longevity and so forth. And while he did not write about non-human bodies, these bipolar efforts to discipline life and promote its health (at least for some beings) can easily be applied to the science and management of animal bodies in industrial settings. Arguably, 'protein' production via intensive animal growing and killing uses the same disciplines to regulate, govern and make the animal bodies efficient and economically viable. Artificial insemination, genetically determined breeding, pharmaceutical inputs, size control for processing ease, feed control for pollution control – these are, in Cary Wolfe's (2013, p.46) words, the 'ur-form' of Foucault's biopolitics in their most 'unchecked, nightmarish effects'.

The capitalist-industrialist model (including state capitalism as practised in Russia and China) emphasises privatisation, profits, economies of scale and 'efficiency' (a term to which we will return). This particular model of production treats animals as commodities. Bob Torres (2007, p.13) calls the animals 'chained commodities'. A commodity is a uniform product traded in the market. The uniformity or standardisation make it easier for people to compare and trade. This, of course, has major ramifications for animal bodies because they must be 'grown' to meet standardised processing requirements, as elaborated in the system of subcontracting in the previous chapter. The various commodities produced from animals have complex, variegated histories – see for example E. Melanie DuPuis's (2002) book on milk, 'nature's perfect food'. All commodities, as Marx and others have pointed out, are the result of social relations involving politics, workers, families, gender, nature, mores, ideologies and consumer subjects. Commodities produced from animal bodies involve even more – the labour of animals, their bodily functions, their social relations (or the disruption thereof) and their deaths.

Changing animals into commodities that are supplied in massive quantities, constantly and cheaply is no small feat and has taken decades of human and scientific engineering, enormous amounts of capital and organisational innovation as well as the reshaping of human attitudes and cultures. Changing an animal with its own desires and social life, its temperamental behaviour and resistance to humans, its own bodily functions and growth patterns into a bundle of commodities that are standardised for a fast-moving market where rapid transport and expected quality are demanded takes considerable effort from multiple parties along the commodity network. Science and technology are required for every step – from breeding to birthing to final processing and rendering. Animal bodies, after meat, are turned into all sorts of other products from leather for shoes, to pet food, biofuel, cosmetic inputs and many more. We cannot possibly sort through all of these for this chapter but we hope to touch on a few.

In a capitalist economy, this all takes place to make profit rather than to produce use value or satisfy an individual or communal need (Gunderson, 2011). Keiko Tanaka and Arunas Juska (2010, p.36) characterise the science and technology of commodity production in agrifood systems as 'politics by other means'. They write:

> By 'politics by other means,' we emphasised the critical role of technoscience, not only in producing agricultural commodities, but also in simultaneously producing and reproducing social structure, including redistribution of wealth, power and status among actors involved in a commodity subsector. By doing so, we tried to raise questions of democracy, public accountability and social justice in designing and implementing institutional changes associated with technoscientific change and innovation.

Their point is that the science intertwined with turning animals into commodities produces outcomes for farmers, their communities, breeding and animal health corporations, the animals, ecologies and consumers. Thus far, the science has yielded greatly increased amounts of meat, eggs, milk and other animal products, but it has also furthered a huge decrease in the number of farmers, a concentration and vertical integration of producers and processors, ecological degradation and questionable health impacts for consumers.

The Political Economy of Animal Science

Jack R. Kloppenburg's (2004, originally published in 1988) seminal work on the political economy of the seed in the United States provides a substantial foundation for animal biotechnologies. He writes about the history of seed commodification, illustrating how ownership and hybridisation made possible the proprietary taking of something that was biologically available to all farmers. A second focus is on the emergence of the division between public and private science wherein the public scientists did the basic science while seed companies did the applied science and selling of the product to farmers. Kloppenburg argues that farmers were not so excited about the scientific establishment and the use of their taxes to develop more complex agricultural methods. Indeed, in the end they were right to be cautious given the numbers of them who were forced out of business over the longer term. Thus, the irony with considering animal science as a means to national development is that livestock science in the past 50 years has driven most farmers out of business. For example, there were 65,000 milk cow operations in the United States in 2009 compared with 97,460 in 2001, a decline of 33 per cent (USDA, 2010). USDA data over the period from 1999 to 2008 show an almost tripling of dairy farm numbers in the very large size (≥2,000 cows) category, from 255 to 730 farms. The percentage of production in this size category increased from 9 per cent to 31 per cent. Data over a longer period show that the average American herd size was 19 cows in 1970, rising to 120 in 2006 (MacDonald and McBride, 2009), with average milk produced per cow doubling and production per farm increasing 12-fold. The larger, more capital- and management-intensive farms tend to be greater technology adopters and have lower relative production costs. They receive higher profits for a while until other farmers catch up or the supply exceeds demand and prices decline. The stage is then set for another technological advance to make reasonable profit margins achievable at reduced production costs. Willard W. Cochrane (1979, p.387), an eminent agricultural economist, called this pattern the 'theory of the treadmill'. For those who never get on the technological treadmill, the results are 'devastating' and they lose their livelihoods either through bankruptcy, foreclosure or other means.

Kloppenburg (2004, p.35) points out that agricultural science is not only a major cause of the technology treadmill that makes some rich and others wage labourers, it is at the same time the producers of the commodities that are 'substituted for the farmer's self-sufficing provision of the means of production'. Removing people from their means of production turns them into commodity consumers and wage labourers, a process Marx identified as 'primitive accumulation'. According to Kloppenburg (2004, p.35),

> new knowledge produced by agricultural science has increasingly reached the farmer not as a public good supplied by the state but in the form of commodities supplied by private enterprises. Agricultural research has greatly facilitated the 'differentiation' of activities off-farm and into industrial settings and can therefore be understood as an essential component of the contemporary dynamic of primitive accumulation in the agricultural sector.

A good example of this, although there are many, is the introduction of recombinant bovine growth hormone by Monsanto, Upjohn, Eli Lilly and American Cyanamid, with the help of university researchers at Pennsylvania State University, the University of Vermont, Cornell University and elsewhere in the United States. Bovine growth hormone or bovine somatotropin is a peptide hormone produced by a cow's pituitary gland, small amounts of which are used in metabolism. Monsanto had tried to develop a process whereby the hormone could be taken from carcasses and injected into living cows but the process was not commercially plausible (Schneider, 1990). When Genentech patented a process for recombinantly producing it in *Escherichia coli* (*E. coli*) in the 1970s, the corporations were off and running towards its commercial development. Injecting dairy cows with this synthetic hormone causes their milk production to increase by several litres per day. At the same time that the companies mentioned were rushing to get the drug or 'input' approved by the US Food and Drug Administration (FDA), there was already a glut of milk. For years there had been an 'oversupply' in the United States and the federal and state governments had intervened to control supply and/or price. In 1986 and 1987 the government paid farmers to kill their cows and stop dairy farming for five years. Some 14,000 farmers participated in this voluntary programme, slaughtering a total of 1.55 million milk cows. One of us remembers seeing the big 'X' on farms in Massachusetts where the cows were to be killed.

Monsanto was the first company to get the drug or 'input' to market and had spent upwards of $300 million to do so (Schneider, 1990). There were many deleterious side effects to the genetically engineered hormone, including cancer risks to humans, more mastitis (udder infection) and other serious health threats for the cows, and early puberty in young people. Cows treated with recombinant bovine somatotropin (rBST) face a nearly 25 per cent increase in the risk of clinical mastitis, a 40 per cent reduction in fertility and 55 per cent increase risk of lameness (Dohoo, et al., 2003). Many studies have noted links between insulin-like growth factor (IGF-1) levels and increased risk of cancer, especially breast and prostate cancer, and rBST increases the level of IGF-1 in humans (Chan, et al., 1998; Holmes, et al., 2002; Yu, et al., 2002). Because Monsanto and the other companies spent so much money on the development of the hormone, they pressured the FDA to approve it quickly, despite reports from other researchers concerned about the side effects (Schneider, 1990). Consumers pressured the FDA to require labelling and organic milk producers sold milk labelled 'rBST free'. Monsanto sued at least one small dairy in Maine because the dairy wanted to label its milk 'rBST free'. Then in 2008 the company reported it was selling the part of its business that produced the drug Posilac. Nevertheless, despite all of the problems, about 15 per cent of US dairy producers use it. A Monsanto spokesperson, Jennifer Garrett (technical services director for Monsanto's dairy business) affirms the effects of the technology treadmill: 'Producing more milk efficiently allows dairy farmers to make more money. The farms with the highest-producing cows are those that are making the most money. Posilac is a product that allows them to do that' (Philipkoski, 2003). Posilac is now owned by Elanco, one of the largest animal chemical companies in the world, which remains intent upon pushing the product in the name of feeding the world (see Raymond, et al., 2009). rBST is not allowed on the market in any of the EU countries or in Canada, Japan, Australia, New Zealand or Israel. USDA organic standards also strictly forbid its use. Whether or not one believes one side or another in the many contests over this productivity enhancer, the bottom line is that it has become a new commodity that has pushed farmers out of business and thinned the ranks of the dairy industry. It also subjects cows to painful, life-threatening conditions – the primary reason that Canada refused it. A dairy farmer outside Ottawa claimed: 'It just puts more pressure on the cows. We don't need it. It's just a gimmick by the makers to get farmers to believe they can get something for nothing. But everything has a cost' (R. Visser, quoted in Bourrie, 1999).

As a national development strategy, the development of animal sciences has helped vertically integrate and concentrate the livestock industry. A good example of this is described by the Kansas City Federal Reserve Bank in a

1991 report. The article states that the 'drive toward contracting and integration in the broiler industry was spurred in the 1950s and 1960s by the need
to keep pace with the high-tech developments of the day – feed formulation,
poultry genetics and mechanization' (Barkema, Drabenstott and Welch, 1991,
p.33). The bank expected that new technologies would encourage a further shift
toward contracting and integration with all of the shifts in rural communities
and other social relations accompanying. The constant drive of scientists to
produce more, bigger and cheaper animals meant that even when there were
political protests against the demise of small farmers, the big companies
continued to produce drugs, feeds and genetics that would yield more from a
single animal or single animal production unit. As Lawrence Busch (2008,
p.149), a prominent sociologist of agriculture, argues, 'most agricultural scientists, overproduction notwithstanding, are still interested in increasing the
production, productivity, and efficiency of farming'.

A Brief History of Animal Science

The Western model of scientific support for agriculture, with the emergence of
institutions fully dedicated to producing new science and technology to increase
and stabilise yields or production, has spread throughout the world. Begun
perhaps with the Royal Agricultural Society of England (formed in 1840) and
furthered in the United States with the Morrill Acts of 1862 and 1890 that
created colleges for educating future farmers (the 'land grant schools') and the
Hatch Act of 1887 which funded research-oriented agricultural experimentation
stations in every land grant college, this model of systematic invention of new
agricultural practices became global after the Second World War (Cochrane,
1979). Louis Pasteur in France and Robert Koch in Germany were bacterio-
logists who identified the cause of anthrax in farm animals as well as developing
laboratory practices for microbiological research that persist today. The Soviet
analyst Rubenstein (1931) characterised the type of science that prevailed to
support capitalism even in the nineteenth century as requiring 'powerful laboratory equipment, intricate, expensive appliances and instruments, experiments
upon a semi-factory scale, a considerable staff for systematic study of the
immense literature growing up on each subject'. He continued:

> In the overwhelming majority of cases, it requires the collective organi-
> sation of labour, the sub-division of the work, and the complex forms of
> co-operation in this work among specialists in various branches of science,
> and of various qualifications. Even when carried on by a large collective
> body, the treatment of many scientific-technical problems takes sometimes
> years, and even tens of years, calling in many cases for tens and hundreds
> of thousands of systematic experiments, tests, and observations. In other
> words, *scientific investigation becomes itself a sort of large scale production*
> organised after the type of industrial plants.
>
> (Rubenstein, 1931, p.1)

Indeed, the institutionalisation of animal science based on the land grant colleges and extension agencies as well as the infrastructures in European countries meant a large number of scientists were focused on issues of disease, herd improvement through breeding and feeding developments. Farmers and ranchers were also involved through their associations. The US Bureau of Animal Industry was created in 1884 to combine the research functions of the Veterinary Division of the USDA with the enforcement functions of the Treasury Cattle Commission. The mandate of Congress to the Bureau was as follows:

> There shall be in the Department of Agriculture a Bureau of Animal Industry. The Secretary of Agriculture is authorized to appoint a chief thereof, who shall be a competent veterinary surgeon, and whose duty it shall be to investigate and report upon the condition of the domestic animals and live poultry of the United States, their protection and use, and also inquire into and report the causes of contagious, infectious, and communicable diseases among them, and the means for the prevention and cure of the same, and to collect such information on these subjects as shall be valuable to the agricultural and commercial interests of the country.
>
> (7 US Code, chapter 15, section 391)

Taking a scientific approach, the Bureau was to examine the causes and effects of animal diseases and the regulatory authority to enforce solutions. Eradicating pleuropneumonia, identifying cattle ticks as the cause of Texas fever, and gaining control over hog cholera were some of Bureau scientists' achievements (Whitaker, 1980).

In 1906 the American Dairy Science Association was formed and the Poultry Science Association began operating in 1908. Researchers in animal nutrition in 1908 formed the American Society of Animal Nutrition. Beltsville, the USDA research facility, began in 1910, and the first meat laboratory was developed in Minnesota. In 1919 the USDA launched a national programme that promoted the use of pure-bred sires to improve American livestock. Pure breeds and inbreeding were thought to be preferable for quite some time until Gregor Mendel's work was rediscovered in the early 1900s and heterosis or hybridising was recognised as a means of instilling vigour and selecting for more desirable traits or breeding out undesirable ones. Selecting superior animals for breeding stock became a specialised industry in the 1920s for poultry, in the 1930s and 1940s for cattle, and in the 1950s and 1960s for pigs (Fuglie, et al., 2011). Companies would invest in pure-bred lines and then sell hybrids to producers. Genetic material was supplied to farmers through pyramid programmes (nucleus herds to multiplier herds to producer stock). The companies could protect intellectual property by guarding access to their pure-bred nucleus herds. This model worked well for pigs and poultry but not as well for cattle because of the greater expense and difficulty of maintaining viable unbred lines.

The development of artificial insemination, a truly international feat, has had a significant impact on livestock genetics, particularly in the dairy industry. Phenomenal growth in the use of artificial insemination in the United States in the 1940s for animals yielded practical methods that spread throughout the world with hundreds of studies published on semen collection, treatment, analysis, insemination, oestrus detection and so on (for a history see Foote, 2002). Both producer cooperatives and private companies have used artificial insemination and performance testing of progeny to improve dairy lines. Sexed semen, primarily for use in dairy breeding, is also available to further the productivity of the industry.

The professionalisation of veterinary medicine, accomplished in Europe and North America during the latter part of the nineteenth century, was established in many other places through colonialism and as a result of raising non-indigenous livestock. International veterinary conferences were held, beginning in the 1860s, and the Office International des Epizooties (OIE) was established in 1924 in Paris to combat European foot-and-mouth disease. Veterinary science (and its broader animal science offspring) has always been intertwined with state building. Livestock epidemics historically threatened sovereign economies and their mediation became a primary function of expanding state bureaucracies (Brown and Gilfoyle, 2010). Expansion of global trade produced new diseases through the dis/replacement of animals in new ecologies and provided an opportunity for concurrent growth of state powers to stop flows, cull and further understand disease aetiologies. Ironically, increased trade produced augmented state power to protect borders of all kinds. Based on the model developed by the Rockefeller Foundation in Mexico in 1920 to eradicate hookworm, the United States promoted the building of institutions for scientific agriculture in many countries.

Certainly in the years since the first agricultural extension services were created, the size of the animal sciences enterprise has altered significantly. Some of the hundreds of journals, sourced from all over the world, that publish animal science include *Meat Science, Animal, Theriogenology, Genetics, Animal Frontiers, American Journal of Animal and Veterinary Sciences, Poultry Science, Journal of Animal Physiology and Animal Nutrition, Chinese Journal of Animal Science, Brazilian Journal of Animal Science, Journal of Applied Animal Science* (from Mahidol University, Thailand), *Revista Brasilerisa de Zootecnia, Reproduction, Fertility and Development*, and *Journal of Pathogens*. A comparison between a Western and Chinese university shows the same foci. The Ohio State University Department of Animal Science lists the following topics as its foci: genetics, tissue biology and processing, physiology, nutrition, bioenergy and nutrient management, and microbiology with a species focus on cattle (beef and dairy), equine, poultry, sheep and pig. The department also includes 'biomass to energy' as a recent addition. The China Agricultural University College of Animal Science and Technology lists animal genetics, breeding and reproduction, animal nutrition and feed science, forage and grassland science. Both universities provide veterinary science education as well. The most recent announcement of the World Conference on Animal Production held in Beijing in the autumn of 2013 lists very similar topics as those identified in the early 1900s in animal science journals in the United States. These topics include animal genetics and breeding, animal nutrition, animal physiology, animal production by type of animal, and finally environment and welfare with the subtopics including only one reference to welfare and that being 'stress and animal welfare' (see World Conference on Animal Production, 2013).

Clearly, agricultural science is big business and those who control it wield enormous power in normalising particular modes of production in the livestock industry. A recent report undertaken by the USDA on global research and development portrays an ever-concentrating structure of research and development with the private sector accounting for about half the total funding (Fuglie, et al., 2011). The authors state that 'a relatively small number of large, multinational firms with global research and development (R&D) account for most R&D in each input industry', with four firms accounting for 50 per cent of global sales in animal health science in 2009. The United States is the world's largest market for animal health products for both food and non-food animals. Companies based in the United States conduct about 42 per cent of private R&D by the global animal health industry. EU countries account for 55 per cent of the R&D with companies in China, Japan, India, Brazil and Israel also making substantive investments.

The development of animal science made possible by private–public complicity has one major outcome – the commodification of animals. Such a commodification of the animal body aims to extract the highest possible economic value out of the animal and in so doing denaturalises any empathetic bonds humans might have with animals. In the following sections, we elaborate on how science commodifies animals and the implications of this for nature.

Fast and Unnatural Commodity Production

Animal sciences break down every part of the animal body to be analysed in terms of its contribution to productivity. As Foucault (1979, p.154) wrote about time, so it is true of the animal body in industrial agriculture: 'the better one disarticulates it by deploying its internal elements under a gaze that supervises them, the more one can accelerate an operation'. The rapid generation of 'improvements' in broiler productivity over the past few decades is testament. Two-kilogram birds that used to be sent to slaughter at four months are now sent at five to six weeks. Persistent selection for rapid growth, high feed utilisation efficiency and large cut yield has led to broilers with superior genetics for productivity. As William Boyd (2001, p.632) writes: 'By the 1960s the broiler had become one of the most intensively researched commodities in U.S. agriculture, while complementary changes in the structure, financing, and organisation of leading firms created an institutional framework for rapidly translating research into commercial gain'. Achieving standards for carcass quality, fat content, *E. coli* presence in carcasses and so forth is very important but the costs must be kept at a minimum by increasing the speed of fruition. The complexities of the work might be illustrated by the title of a recent article: 'Effects of N,N-dimethylglycine sodium salt on apparent digestibility, vitamin E absorption, and serum proteins in broiler chickens fed a high- or low-fat diet' (Prola, et al., 2013). N,N-dimethylglycine (DMG) is used as an athletic performance enhancer in humans and horses. It purportedly improves nutrient digestion and reduces nitrogen emissions. The study mentioned above is designed to deal with ascites in broilers – a condition caused by heart failure due to the inability of lung capacity to keep up with growth of muscle. The chicken is growing too fast for his or her lungs which grow slower than the rest of its body in broiler production. The bird does not get enough oxygen and may fill with fluid or get cirrhosis of the liver. Of course, mouldy litter and ammonia-filled air also cause ascites in poultry. The adding of DMG to feed would purportedly mediate this problem by allowing more oxidation. This example is emblematic of hundreds of issues that arise from this model of industrial-scale production at rapid rates of growth.

This exploitation of genetics, feed, housing and management for maximal growth though has led to undesirable side effects such as heart failure and leg problems. Other disease problems from close housing as well as litter-induced foot and breast burning also occur. These problems generate a further research agenda for animal scientists all over the world. Now that there is a market for chicken feet in China, researchers are trying to ascertain how to avoid foot burning (or footpad dermatitis) as it disqualifies broiler feet for those markets. A footpad scoring system has been developed to allow calculation of the extent of the problem for sales and welfare.

This speeding up of livestock commodity production includes milk, eggs and red meat as well. Dairy cow milk production has quadrupled over the past few decades but the cows are 'spent' at a much younger age. Cows can live as long as 20 years but their time on the 'farm' has decreased to an average of four in the United States. They are bred younger and more often with artificial insemination and the new technologies that allow for the identification of oestrus. Pigs also are grown as fast as possible – usually slaughtered at 22 weeks. As one feed supplement marketer says: 'The profitability of a pig farm is based upon getting as many pigs to saleable weight within the shortest possible time. The faster a pig reaches its saleable weight the less feed is consumed and the quicker the return on the boar insemination investment'. For cattle, feedlots do the job of getting cattle to market sooner. Pasture-fed cattle take a longer time to gain the weight necessary to make them optimally slaughterable. Seventy-five years ago, steers were four or five years old at slaughter; now they are 14 or 16 months. To take a 45 kg calf to 550 kg in a year takes quantities of corn or soybeans, protein supplements, antibiotics and other drugs, including growth hormones. Hormones are of double benefit because the cattle gain more weight and do not require as much feed. Antibiotics also meant that animals reached market weight faster when fed as a feed additive. It is worth noting that growth hormones are banned for poultry and pigs in the United States and for all animals in Brazil. The EU banned most of them and the controversy continues to effect trade and tariffs.

The genetics revolution will contribute to the speed of commodity production through the decreased time necessary to evaluate breeding values. DNA marker technologies and high-throughput SNP genotyping technologies, accompanied by reduced costs for genotyping and sequencing, make genetic selection possible, thus accelerating genetic gain by reducing generation intervals. High-throughput technology allows for the inheritance of hundreds of thousands of markers to be traced through generations at about approximately $200 per animal. The breeding industry is adapting selection procedures in each species to include this tool. Genetic estimated breeding values would be substituted for estimated breeding values which rely on generations of offspring to prove. In dairy cattle breeding programmes, genomic selection allows breeders to identify genetically superior animals at an earlier age – even before sexual maturity. Once these genetically superior animals are identified using genetic testing, they can become part of a multiple ovulation and embryo transfer (MOET) programme. Generally, a cow may produce only one calf per year. But with MOET, a genetically elite cow can produce many cows per year. Hormones are given to the cow to produce more eggs than the one that is generally produced, and then the cow is artificially inseminated with semen from an elite bull. A few days later, the multiple embryos are flushed out through catheterisation and then placed in other calves at the correct stage of estrus. Genomic selection is done for high yields, for improving the fatty acid and protein profile of milk, among other interventions. The animal breeding industry, which is now global, is adapting selection procedures in each species to include this new technology.

Currently, there is considerable emphasis on yield per animal, a reflection of expensive genetics and the cost of feeding. Although many producers are still adding animals or raising as many animals as possible, others have focused upon optimising the productivity of each animal body. For example, US beef processors will give discounts to producers for beef carcasses over 450 kg. These beef cattle are substantially larger than were those marketed during the mid- to late twentieth century in the United States; the average weight of a fattened steer sold to a packing plant is now roughly 600 kg – up from 450 kg in 1975. Pigs are also subject to this newer focus on yield per animal body. Reproductive performance is most critical in pigs managed in confinement because of the large capital investment in facilities. Smithfield, for example, implemented a new sow contract in 2011 which rewards heavier pigs weaned at an older age, increasing sow productivity and a consistent flow of pigs (Freese, 2011). The expectation is that sow inventory levels will continue to decline as increased sow productivity allows fewer of them to provide the necessary market pigs. Higher-performing pigs are expected to increase daily weight gain and decrease in mortality and cull rates. Danish pig producers are aiming for 35 weaned pigs per sow per year, up from 30 in 2010 (Lumb, 2010). The 'target', as they call it, was only 25 weaned pigs per sow in 1995.

The goal again is 'improving efficiency and economical output'. Over the past couple of decades sow length has increased by 10 per cent and weight by 20 per cent, but they have not developed more teats. Therefore, the extra piglets have to be fostered to other sow mothers. Of course the generation of more weaned pigs per year puts more stress on sows and requires more research. Feet and legs are important because sows are expected to farrow more than two litters per year, nurse a large litter of pigs for two to three weeks, breed back in seven days or less, and live their entire lives on solid concrete or wire floors.

These livestock breeds developed for high input production systems are of ever more special genetics and are expensive to buy and develop. For example, a top-of-the-line gilt (unbred sow) might go for more than $2,500 (a straw of semen from a superior sire may cost more than $1,300). She may not breed for three months while she is being fed so there are 'development' costs. Then, there is a downward U-shaped curve to the number of her piglets per litter as she goes through multiple births. The weight of piglets per litter increases with her 'parity' (birthing event) until a decline at the end as well. A sow has to remain in the herd for multiple parities in order to generate adequate profit to cover the initial gilt purchase price plus the associated gilt development costs. This statement reflects a more individually focused approach given the high cost of special genetics.

Animal welfare, public health and environmental sustainability goals add even more pressure to this cycle of research, which Adrian Smith, Andy Stirling and Frans Berkhout (2005) call 'endogenous renewal' – the efforts of all members of a supply chain attempting to find ways of responding to threats to the existing regime. For example, one of the threats to the pig industry was the public concern with high-fat meat. Scientists and other industry actors responded with the leaner pig, but breeding for leanness and heavy muscling generated problems with hypertension, tremors, sudden death and infertility or porcine stress symptom (PSS). PSS (related to malignant hyperthermia in humans) has been identified as associated with a candidate gene called the ryanodine receptor. The development of a DNA-based test for the gene allows breeders to get rid of any pigs that would be candidates for the disease or to use the gene to breed in leanness and muscling. More recent Brazilian research illustrates just how much more acidic and poor-quality pork might be from pigs with the gene for PSS (Band, et al., 2005).

'Euthanasia' Experiments and Rendered Commodities

The story about Henry Ford getting his idea for vehicle assembly from the Chicago slaughterhouse disassembly process is common currency. Just like every other part of the animal commodity production process, however, there is considerable research – on how to kill these living beings in mass quantities both on and off farm. For example, an article in *Poultry Science* in 2004 is entitled 'On-farm euthanasia of broiler chickens: effects of different gas mixtures on behavior and brain activity'. These researchers tested different gas mixtures on two- and six-week-old chicks. Here is what they observed: 'In all 3 gas mixtures, head shaking, gasping, and convulsions were observed before loss of posture. Loss of posture and suppression of electrical activity of the brain (n = 7) occurred almost simultaneously' (Gerritzen, et al., 2004, p.1294). Ironically, these studies (and there are many of them) are done in the name of 'welfare' because CO_2-caused death, while sanctioned by the American Veterinary Medical Association (AVMA), is painful because CO_2 is acidic and the animals do not die right away.

Maceration is another commonly used way of killing day-old male chicks and in-egg chicks. This is basically throwing them in a grinder-crusher machine while still alive. The Federation of Animal Science Societies considers it a reasonable way of killing poultry and there are several companies which produce the equipment. Agriculture Canada, the AVMA and the French Ministry of Agriculture apparently agree. For example, the Breuil and New Tech Solutions combined roller system, produced in Gainesville, Georgia (USA), is designed for one-day-old and 'live in-shell' chicks and its documentation says it provides for both crushing and tearing. Because of complaints by animal welfare groups and due to the waste disposal problem (over 260 million male chicks are culled annually in the United States alone) as well as the extra costs (eggs are all vaccinated), the layer hen industry is under pressure to find other ways to deal with the unwanted male chicks. Research on using infrared spectroscopic imaging to identify pre-incubation sex is underway as are other experiments to change or control sex, or to produce a killing gene for males.

As elaborated in greater detail in Chapter 5, Temple Grandin, the famous icon of industrial slaughterhouse design, has written widely on various stunning techniques for large animals. Her approach is to produce environments and management systems wherein animals will willingly enter stunning or gassing areas – using visual techniques and air blowing in the animal's face among other things. An example of her work on stunning is provided here:

> To produce instantaneous, painless unconsciousness, sufficient amperage (current) must pass through the animal's brain to induce an epileptic seizure. Insufficient amperage or a current path that fails to go through the brain will be painful for the animal. It will feel a large electric shock or heart attack symptoms, even though it may be paralysed and unable to move. When electric stunning is done correctly, the animal will feel nothing.
>
> (Grandin, 2013)

Her goal is to minimise the need for rough handling, electrical prods and other abusive practices. Newborn calves can hardly walk into the slaughter line and seem to receive an extra amount of rough handling. Grandin says she has seen them thrown, dragged by ears and otherwise abused by their handlers. For example, an organic slaughterhouse in Vermont (USA) was temporarily closed (for the third time) because workers were kicking, cutting and shocking and dragging the male calves (dairies do not require males).

Grandin has also found a number of problems with ractopamine- and beta-agonist-fed livestock who come into the slaughterhouses lame and unwilling or too weak to walk without abusive prodding. She has been a pioneer in using vocalisation as a measure of animal welfare in slaughterhouses.

There is also research done on getting the animals out of their cages and pens and to the slaughterhouses. A study undertaken in Britain of spent hens from battery cages found that nearly 30 per cent of them had broken bones by the time they got to the water bath stunner (Kristensen, Berry and Tinker, 2001). Laying hens have bone problems because they are forced to lay so many more eggs than would be natural – all of which takes away calcium from their bones. Then when they are 'spent', usually after one year, they are manually taken from their cages, carried upside down by their legs, passed from one hand to another – often with three to five other hens – and then passed off to other handlers before entering the cages that will take them to the killing line. So people actually study how to avoid the high number of broken bones – but they are also concerned about the labour welfare as well.

How to dispose of spent hens is another task for science because their previous rendering for chicken feed has become somewhat unpopular following the incidence of bovine spongiform encephalopathy (BSE, mad cow disease) in Britain. In Sonoma County, California alone some half a million birds a year have to be disposed of because they are not productive enough and the market for spent hen meat has collapsed. After gassing them with carbon monoxide, their carcasses are put in giant compost heaps. Every now and then one or more stagger out of the pile – the neighbours call them 'zombie chickens' (Young, 2006, p.1). In the United States there are annually more than 250 million spent hens to dispose of, and because of their brittle and broken bones and their often emaciated state they are now considered unsafe for school lunch programmes, once a primary market. They are also more likely to be contaminated with *Salmonella* than are broilers. The breakdown of their immune system, which may happen with forced moulting caused by three to 14 days of starvation (in order to produce another cycle of egg laying), causes them to be more subject to bacterial infection and other disease. In an effort to evaluate the compost potential of spent hens, the Alberta Agriculture and Rural Development Ministry (Canada) laid 5,800 spent hens along windrows of fields and covered them with dirt and wood shavings to test composting speed and quality (polluted run-off is of course a problem) (Young, 2006). In another enterprising move, the Victorian Farmers Federation Egg Group (Australia) worked with a biofuel company to study the feasibility of building a plant to turn spent hens into biofuel (Berkhout, 2010). Even more imaginative is the research on using spent hen meat in place of Alaskan pollack to make imitation artificial crab sticks (Jin, et al., 2011). Alternatively, they can be rendered into slurry, and be made into poultry by-product meal, which can be incorporated into animal feeds.

'Depopulation' is the term most commonly used for mass disposal of animals threatened with or carrying diseases problematic to humans (that is, foot-and-mouth disease, various flus, BSE). Following the outbreak of foot-and-mouth disease in Taiwan where 5 million carcasses required disposal, and in Britain where 2 million animals from 18,000 farms were killed and disposed of at a total cost of £2.8 billion (Anderson, 2002), hundreds of studies were done on how to undertake such mass killings and disposals. The state of Virginia (USA) has had to deal with two cases of avian influenza that required the killing and disposal of massive numbers of birds; in 1984 over 5,700 tons of poultry carcasses required disposal and another 16,900 tons were disposed of in 2002. Researchers at the University of Delaware have done work on using foam and the US government has approved the use of firefighting foam to kill birds in case of an avian flu outbreak (Dawson, et al., 2006).

Incineration, anaerobic digestion, composting, landfilling, alkaline hydrolysis, grinding, ocean disposal, freezing – even napalm and feeding to alligators – have been studied and evaluated as possibilities for disposing of large tonnages of animals in concentrated animal feed operations (CAFOs) (McClaskey, 2004). The Food and Agriculture Organization (2011) estimates that half of each cow and over a third of each pig and poultry animal are not used by humans. Considering that billions of each are slaughtered each year, there is a lot of leftover animal biological material requiring disposal. The science of rendering aids corporations in their efforts to make profits on this dead matter. In most rendering systems, the materials are cooked, then pressed and treated with other measures to separate the fats from the proteins (blood meal, bone meal and poultry meal). Blood from slaughterhouses is also processed. For example, spent hens are being evaluated at the University of Alberta as a potential source of adhesive (glue) – one litre can be made from one carcass (Willerton and Proulx, 2012).

In the United States, the Fats and Proteins Research Foundation (FPRF) does most of the research for the National Renderers Association. Pet food and energy are primary uses of rendered bodies. The FPRF's research on the fuel's burning characteristics and emissions has allowed for the permitting of substituting animal fats for no. 2 or no. 6 fuel oil or natural gas for the production of steam. Carbonised chicken feathers are being tested for feasibility as hydrogen storage opportunities in mobile transport. Researchers at the University of Delaware are doing some of this research as well as evaluating chicken feathers as sources for bio-based computer circuit boards and for clothing. This latter research is touted as 'green chemistry' and is winning awards; the problem is that it may even further lock in entirely unsustainable and cruel structures of production under the auspices of being 'green'.

The work of finding more uses for these leftover parts has increased since BSE hit Britain. Prior to that, nearly all of the material was used for animal feed. But now many countries do not allow any mammalian animal parts to be re-fed to animals. In the EU mammalian meat and bone meal are banned from any feed used to feed animals for human consumption. The United States does allow it though with the exception of feeding ruminant brain and spinal materials to ruminants. Even hydrolysed feathers are fed as protein (called 'feather meal') and considerable research has gone into ascertaining how to process them for maximum digestibility (Moritz and Latshaw, 2001).

However, unbeknown to most consumers, the drive towards commodification and standardisation has more insidious biopolitical impacts on food animals, beyond regulating their relative outward characteristics of weight, length of growth and body fat content. One such impact relates to the notion of 'docility'. In *Discipline and Punish*, Foucault (1979, p.136) writes about the docile body which 'joins the analyzable body to the manipulable body. A body is docile that may be subjected, used, transformed and improved'. He was discussing the human body and in particular the body of the soldier, but his analysis applies very directly to food animals, particularly since the 1950s when confinement operations and genetically guided breeding became popular. Of particular interest to us is what he called the 'modality' – the 'uninterrupted, constant coercion, supervising the processes of the activity rather than its result ... exercised according to a codification that partitions as closely as possible time, space, movement' (Foucault, 1979, p.137). He distinguishes these regimens and disciplines of the body from slavery and service, however, which we contend are states within which food animal bodies in CAFOs are maintained. He argues that

> what was then being formed as a policy of coercions that act upon the body, a calculated manipulation of its elements, its gestures, its behavior. The human body was entering a machinery of power that explores it, breaks it down and rearranges it. A 'political anatomy', which was also a 'mechanics of power', was being born, not only so that they may do what one wishes, but so that they may operate as one wishes, with the techniques, the speed and the efficiency that one determines.
>
> (Foucault, 1979, p.138)

In the past 50 years, the calculations of animal body productivity and potential have been sharpened and rationalised. There are several often conflicting goals for the scientist working on engineering animals for processing. Breeding in both pigs and cattle for leanness and productivity has generated more flightiness and, in pigs, tail biting. Daily movement of thousands of terrified animals through slaughterhouses undoubtedly calls for research on behaviour. Bruising and lacerations, as well as elevated cortisol levels, can ruin meat quality. Pigs that are highly stressed can end up with 'pale soft exudative' tissue or dark dry meat (as can also poultry, cattle and sheep). Researchers have thus established a measurement of 'docility' that seeks to capture this state of affairs. Poor docility, or fear of humans, is actually a survival trait in the wild. In domesticated animals, particularly large animals like pigs and cattle, such flightiness is undesirable and can adversely affect productivity, net feed intake and meat quality. Docile animals are less likely to stress during handling and transportation, and is thus important to many people – not only ranchers and farmers, but feedlot workers, transporters, breeders, stock agents and slaughterhouse processors. There are multiple tests for docility, primarily based upon animal reaction to humans (and each other) in variously sized enclosures. According to the industry, docility scoring is important in pigs, especially in loose housing.

Mostly, animals that are aggressive are culled (removed from the herd either through death or sale). This is not only a factor of safety for the workers and handlers but also for the other animals. Animals that are considered too docile are problematic as well because they are unable to compete for food or protect their young adequately. Science is deemed necessary to determine how to breed animals (or select breeds) that are optimally docile for particular environments. Docility scores generally go from one to five or one to six, with a one being 'docile' and the highest number being 'aggressive'; it is an effort to make food animals easier to handle and process. Cattle tests are usually done by measuring flight speed or in the crush (the narrow groove through which animals are held during inoculations or dehorning), the pen or the yard. Each breed has its own rating system for cattle and the breed organisation keeps the scores. A pig test developed from the chute test for cattle includes the following measurement for activity when being weighed and having back fat measured: 1) remains calm with little or no movement; 2) walks forwards and backwards at a slow pace; 3) continuously moves forwards and backwards at a rapid pace; 4) continuously moves forwards and backwards at a rapid pace, with high-pitched vocalisation; and 5) continuously moves forwards and backwards, vocalises and attempts to escape by jumping or digging. Piglets are also evaluated using the 'back test' in which they are placed on their backs in a supine position and timed for 60 seconds to see how much they struggle and how many efforts are made to escape the position. Yet another test for aggression or docility is the 'resident intruder test' in which individual piglets from two different litters are placed in the same closed-off space and watched for up to five minutes.

The behaviours these tests measure are considered moderately heritable but that conclusion is contested. Many people insist that docility is inherited while others, mostly in the small farmer league, insist that you can turn just about any pig into one who likes belly rubs – that it is nearly always learned. With more people raising animals themselves and educating each other about the 'natures' of different animals, we will learn more about the breed-specific nature of animals vis-à-vis their environment or management. Nonetheless, animals in industrial settings cannot be too docile or they may not be good mothers or assertive eaters. Prolific layers are purportedly more excitable than other types of hens. It is about fine-tuning a balance and finding those sensitivities that can be controlled by breeding, feeding, lighting, spacing and other practices that might slightly enhance productivity. The new technologies breed contradictions which must then be solved for the constant and dynamic pursuit of optimality.

Despite the hyper-rationalisation, standardisation and specialisation in the livestock industry placing pressure on animals to conform and behave, not all the life and vitality can be eliminated. In fact, as Anna Williams (2004) points out, animal sentience is enrolled to get crowd control over stressed, frightened animals. She recounts the use of decoy animals to get others to peacefully enter slaughterhouse lines. She also describes the way Temple Grandin's engineering of slaughterhouse line architecture uses animal tendencies to get them to willingly go to their deaths. But there are still those who refuse and run. On 10 April 2012 a 340-kg black-and-white cow escaped from the slaughterhouse and ran through Patterson, New Jersey until tranquillised by somewhat hysterical policemen. The cow was taken to Woodstock Farm Sanctuary in New York. Another cow ran away from a slaughterhouse in Queens in August 2012 but did not get rescued quite as nicely. Then there was 'Molly' from Montana who also escaped a slaughterhouse in 2006. The science, capital and institutions that attempt to commodify food animals for the purported good of the consumers will continue to find new ways to intervene with the biological lives of these animals. In so doing, and coupled with the large producers doing whatever they can to obscure the realities of intensified farming, a new normal is presented. This new normal is materialised through an exemplary governmentality that shapes the *biopolitical links* between food animals and human consumers.

Conclusion

The global industrial meat complex is far-reaching in the way it transforms the techniques of production, yet its main contours can be traced. Politics, governing institutions and the economy (highlighted in the previous chapter) play critical roles in the commodification of the animal, as is the role of science and technology detailed in this chapter. While this complex extends its grip unevenly across the world and impacts on local livestock industries differentially, what remains unambiguous is its unilateral goal of extracting more from each animal through commodification and intensification.

The more the animal becomes commodified and commoditised, the less the animal science enterprise (and subsequently the average consumer) treats the animal as a being with social behaviour, desires, relationships and emotions. Kloppenburg (2004) writes about the reductionism that takes place when the seed is taken apart and worked on in genetic bits. It ceases to be a whole seed and instead becomes a series of objects to which property rights can be extended. Farm animals became packages of molecules, genes, muscles, fats, enzymes, hormones, reproductive organs and carcass qualities. With the introduction of artificial insemination in the 1940s, along with in vitro fertilisation, many animals no longer had to have physical contact to reproduce. Cloning of course makes that even more the case. Incubated chicks no longer had mothers, mothers no longer had chicks. Dairy cows never touched their offspring and pigs were weaned well before their mother's milk dried up.

The animal science enterprise treats animals as flesh, milk and egg machines – replaceable parts in a parade of breathtaking magnitude to slaughter. Trillions and more – can we even count? The science is ultra rational, studying subjects such as how to break social bonds without decreasing production levels and the language of science becomes the new normal in our treatment of food animals. Animal bodies are 'machines made of meat' – stimulus–response mechanisms whose behaviour must be moulded to fit into a commodity. From Alexander Cockburn's (1996, p.20) short history of meat:

> 'The breeding sow', an executive from Wall's Meat Co. wrote in National Hog Farmer in the late 1970s, 'should be thought of, and treated as, a valuable piece of machinery whose function is to pump out baby pigs like a sausage machine.'

In the days before industrialisation, on the other hand, chickens were part of rural family life. In her history of the development of poultry farming in northern Georgia, Monica Richmond Gisolfi (2006, p.177) quotes the daughter of a tenant farmer who remembers that 'her family always had a yard flock', that chickens were 'a part of life all the way'. Contemporary urban chicken keepers talk about how some of their chickens are friendlier than others and how brave mother hens can be when their chicks are threatened. In many parts of the world, urban chicken farming has persisted and, indeed, flourished (Hovorka, 2006). A pig farmer writing about the wonders of the Berkshire breed in 1920 commented on what 'kind, careful' mothers Berkshire pigs are. Smaller producers who know their animals may tend to speak of their animals as living beings with personalities, but the large industrial producers do not consider individual animals as they are dealing with tens of thousands. Many of these industrial systems are nearly 100 per cent automated so there is very little contact between humans and animals. But we should not get too romantic about smaller producers. Many small dairy farmers separate male calves at birth from their mothers – one outside Worcester, Massachusetts, cautioned us not to be 'sentimental' about the fact that the calves do not even touch their mothers (or their mothers them) because they might damage the udders. These calves, fed milk substitutes or poorer-quality milk, are sold on to slaughterhouses that process veal. Just this past year in Vermont an organic dairy was closed down for cruelty to one-day-old calves that could barely walk (they have to be able to walk into the slaughter line or they are not supposed to be used).

The commodified livestock body is colonised through discursive and management practices that for the most part ignore the vitality and the social and emotional life of the animal. The body's metabolism is deployed, reshaped and prodded to produce more, faster and cheaper. Social and familial bonds are severed or ignored, instinctual desires to forage and root are thwarted, and intelligent animals are left to fend for themselves in barren and uncomfortable spaces – often too small for much movement. Drugs are administered that promote growth of specific kinds which in the end debilitate the animal and leave them sick as they enter slaughter lines. The technologies and sciences that have made this possible have undercut the production potential of millions of farmers and ranchers around the world, rendering industrial livestock production a concentrated and integrated industry. Only the beef, goat and sheep producers are still to some extent free of contracts and other oligopolistic pressures, although this is also changing.

The commodification of animals through these enjoined practices of finance and science, aided in many places by governments, has provided surplus capital to the few at the cost of the many. In sum, it has produced a new and radical governmentality of food animals which obscures the externalities of the industry as it becomes more concentrated to the few that wield increased economies of scale and power. Such externalities impact severely on animals and ecologies alike as well as adjacent human communities. In the following chapters, we proceed to look more closely at the impacts of the intensification and commodification (in terms of the environment, people and animals) and some of the mitigation efforts undertaken by both the very actors that produce such impacts as well as activist groups who continue to rage against the global industrial meat complex.

4 The Global Meat Factory and the Environment

Introduction

Across the world more resources have been devoted to producing meat. Greenpeace, in its long-standing fight against deforestation in the Amazon, states that, contrary to popular belief, it is the cattle sector and not the timber industry that is the single largest driver of global deforestation. The production of beef is responsible for 14 per cent of world's annual deforestation and 80 per cent of all deforestation in the Amazon (Greenpeace, 2009). In a comprehensive critique, the FAO's 2006 landmark study, *Livestock's long shadow*, unleashed a litany of environmental impacts caused directly or indirectly by the meat industry. It is revealed among other things that the livestock sector releases 18 per cent of greenhouse gas emissions (measured in CO_2 equivalent), which is more than the entire transportation sector (Steinfeld, et al., 2006).

The global intensified livestock industry not only has clear political-economic roots, it bears distinct ethical-environmental ramifications. In Chapter 3, through the lens of biopolitics, we discussed how science and technology play a critical role in the commodification of food animals. Such commodification in turn is underpinned by a political economy that (re)produces ecological impacts with apparent impunity and the complicity of governing institutions. The multifarious consequences of the global livestock industry can in turn be viewed through the broad lenses of political ecology, for it is an industry that precisely 'combines the concerns of ecology and a broadly defined political economy' (Blaikie and Brookfield, 1987, p.17). The world is now firmly in the midst of a 'brown' revolution. Farmers produced 276 million tons of chicken, pork, beef and other types of meats in 2007, four times more than in 1961 (Halweil and Nierenberg, 2008, p.61). This figure has since risen to 308 million tons in 2013 (Heinrich Böll Foundation, 2014), a more than 10 per cent increase in just six years. As elaborated, such a dramatic growth is largely due to the unceasing intensification of the modern meat industry. For example, in the United States, although the number of pig farms had decreased significantly from 2 million in 1950 to 73,600 in 2005, the production of pigs in the same period rose from 80 million to 100 million. This trend is mirrored in the slaughter industry as well. Despite the dramatic increase in food animals that needed to be slaughtered, the number of slaughterhouses in the United States actually fell from 10,000 in 1967 to 3,000 in 2010 (Heinrich Böll Foundation, 2014, p.14). This intensification process is accompanied by other key political-economic developments in the livestock industry, as mentioned in Chapter 2. First, the production of meat has increasingly relied on contract farming where different farms are contracted, by larger meatpacking companies, to rear livestock at specific stages of the animals' growth. Second, there has been a growing market consolidation of the top meat producers, particularly in developed economies, through expansion, mergers and acquisitions. For example, just 10 companies processed 88 per cent of the total number of pigs in 2012 (Heinrich Böll Foundation, 2014). As mentioned earlier, the top 10 global meat companies have become massive revenue-generating conglomerates.

In the previous chapter, we have seen how the control of the meat sector by a few extends to science and technology as well. In turn, science and technology also make possible the mechanics of the intensification of meat production. These inputs and changes to the global trend have been devastating to food animals, the environment and communities of people. In this chapter, we focus on the latter, leaving the specific question of animal welfare to the next chapter. While we discuss the gamut of environmental and human impacts of industrial farming, we also highlight and access the attempts to mitigate some of these impacts. We argue that ironically the mitigation process is imperative to sustain, uninterrupted, the continual commodification of food animals and its subsequent normalisation. We hence question if these solutions go to the root of the issue rather than merely whitewashing its surface.

The Intensification of the Global Livestock Industry

The top three most popular meats in the world, by tons consumed, are (in descending order) pork, poultry and beef. Together, they represent 93 per cent of global meat output. Such a dramatic increase in demand and supply is not possible without concomitant changes in the way meat is produced. The defining feature of the contemporary meat industry is its unceasing concentration and intensification – fewer but bigger farms or factories, with more specialisation of feed and other inputs, and fewer farm workers. The United States is the leader in the production of meat (especially poultry and pork) in terms of technological advancements and organisational innovation. Several leading companies now control most of the supply of meat in the country. In 2005 the top three beef packers in the United States controlled more than 80 per cent of the market while the pork packing industry was 64 per cent controlled by four companies, up from 40 per cent in 1990 (Hendrickson and Heffernan, 2007). Some of these leading meat companies include Tyson Foods which saw sales of $40.6 billion in the 2015 financial year (Tyson Foods, 2016); and Smithfield Foods which netted sales of $15 billion in 2014 (Smithfield Foods, 2014). More generally, food conglomerates have continually achieved high profits in the midst of global hunger, as have the top fertiliser corporations in the world, including industry giants such as PotashCorp (Canada), Mosaic (USA), Yara (Norway), Agrium (Canada), CF Industries (USA) and Sinochem (China).

For leading meat companies and other investment firms too, the search for bigger markets to exploit and bigger profits to reap has extended beyond national boundaries in the past decade. As mentioned earlier, the Brazilian beef-packing company JBS has, since 2007, bought out several leading meat-packing companies in the United States, including Swift & Company, and is now the biggest beef processor in the world. India, Australia, Brazil and the United States each exported more than 1 million metric tons of beef in 2015, while India and Brazil have the two largest cattle inventories in the world. In 2007 Goldman Sachs bought out the largest pig producer in China for $252 million, only to sell its stake barely three years later. China's case is particularly instructive. Although China is the world's biggest consumer of pork and accounts for 45–48 per cent of the world's pork production each year (*Pig International*, 2007, p.12), its pig industry is relatively undeveloped. Data released by China's Ministry of Agriculture in 2002 put the number of pig 'farms' in China at an astounding 105,367,514. For the most part, such farms produced just two or three pigs for sale each year and are more akin to subsistence farming. As recently as 2002, only 4,132 farms produced more than 3,000 pigs a year (*Pig International*, 2005, p.11). By 2005 farms that produced fewer than 100 pigs a year still made up 70 per cent of total output (Yin, 2006, p.22). While China represents a gold mine for global meat processors to capture, the pressures of supplying sufficient pork to its citizens means that China has adopted a two-prong strategy of modernising its domestic production as well as facilitating Chinese companies to take over established pork processors in other countries, as noted below.

Besides China, other post-socialist countries in Central and Eastern Europe which are transitioning to a more market-based economy offer immense possibilities for investment. Smithfield, the largest pig-producing company in the world, has made several investments in Eastern Europe in the past 20 years. In fact, Smithfield can be found in countries such as Spain, Poland, Romania, Britain and Mexico and, as mentioned earlier, the company was itself bought over by a Chinese company in 2013. The spread of these global meat producers has meant that more meat is being produced by fewer farms in fewer places. This sometimes results in the daily transportation of millions of animals from one end of the world to another – an unsurprising phenomenon because, for the most part, many countries are not self-sufficient in meeting their demand for meat. Beyond that, as Schneider and Sharma (2014, p.32) succinctly point out in reference to the Chinese pig industry:

> Chinese policymakers see the U.S. industrial pork production model as the solution to China's food safety problems. Their incentives to consolidate, scale up and standardize the industry are intended to increase

production efficiencies and reduce food safety scandals, Yet it is precisely this system of factory farming (developed in the U.S. over 50 years ago) that confines large numbers of animals together in restricted spaces and creates systems for fast animal growth and short meat-to-market times that has led to drastic environmental, public health and animal welfare problems in the U.S.

The changes in the intensity of the livestock sector, reflecting the economic logic of a Fordist regime, thus produce significant social-political and environmental ramifications. In such a regime, production tasks are divided into minute detail and goods are mass-produced. For the livestock industry, mass production has led to environmental consequences at scales that were unheard of as recently as 30 years ago. The production of meat has continued to be predicated upon increasing productivity and standardisation. For ease of transportation, slaughtering, packaging and consumers' perennial demand for health and convenience, livestock animals in 'modern farms' are reared to precise requirements (Ufkes, 1998). In many cases, the animals are owned by the processors or slaughterhouse owners from birth to death – contract farmers are essentially hired hands. Increasing numbers of farmers are adapting to such 'modern' production and organisational methods (for example, a contract system of farming and highly mechanised and circumscribed modes of production) introduced by established meat companies. With the complicity of public institutions and private capital interests, farmers are also more likely to accept such changes as beneficial to themselves given, for example, the proclivity of the 'market' for 'standardised meat'. Commodification and intensification are thus presented as a 'natural', good development and essential for survival to many less-developed livestock producers.

Intensification processes also involve the 'economic colonisation of the rural areas' and it is a colonisation which should be resisted by rural people so as to 'preserve their priceless rural culture ... and to pursue a different strategy of sustainable rural development' (Ikerd, 2003, pp.34–8; see also Whatmore, 1994 for a European perspective). The anthropologist Walter Goldschmidt concurs. Reflecting on the intensification and concentration of the rural pig industry in Iowa, he notes that the 'sense of community, the ideals of mutuality and the social value of civility' are eroded by the changing systems of production (Goldschmidt, 1998, p.185). Further, the welfare of the workers employed in big factory farms is another concern, as is the plight of animals in the midst of such economic transformations (see Chapter 5).

The Ecological Impacts of Livestock Production and Consumption

Industrial livestock production is one of the most significant generators of ecological impacts on global, regional and local scales (see Emel and Neo, 2011 for more discussion). Flooding the global markets with cheap meat, milk and eggs has huge implications for biogeochemical cycles and land cover change. Some 25 per cent of the earth's surface is managed grazing, making it the biggest category of land use (Asner, et al., 2004). Thirty-four per cent of the world's cropland is dedicated to producing feed for livestock. Counting grazing land as well as lands in feed crop production, the livestock sector occupies 30 per cent of the ice-free terrestrial surface of the planet (Steinfeld, et al., 2006, p.4). Transformation of forest and grassland into rangelands and fodder or grain crops is occurring at alarming rates, especially in South America (McAlpine, et al., 2007). Pastures and feed crops account for a 70 per cent decrease in forested land in the Amazon (Steinfeld, et al., 2006). These land use changes generate CO_2 emissions, alter biodiversity and hydrologic cycles, and produce new pollutants. Concentrated livestock production generates its own set of hazards to people and the environment, including serious nitrogen and phosphorous pollution of water resources, new viruses and exotic new drug-laced pollutants. The FAO's comprehensive study of the ecological effects of livestock production globally generated considerable press commentary because the authors found global livestock production responsible for 18 per cent of greenhouse gas emissions, making it the number one anthropogenic source (Steinfeld, et al., 2006). The Pew Commission followed the FAO's groundbreaking study with its own expert-led examination of the industry in the United States. Its conclusion: 'The present system of producing food animals in the United States is not sustainable and presents an unacceptable level of risk to public health and damage to the environment, as well as unnecessary harm to the animals we raise for food' (Pew Commission on Industrial Farm Animal Production, 2009, p.viii).

The primary types of ecological impacts we will consider here include greenhouse gas (GHG) emissions, water pollution and supply issues, air pollution and biodiversity impacts. These impacts differ according to the type of livestock considered, the type of production system involved (extensive or intensive), and the geography of the production sites (rural–urban, arid–temperate–tropical, coastal–interior and so forth). While we will not be able to consider all of the impacts, we briefly examine the broad impacts that are detailed in other sources. We discuss the implications of these impacts and briefly consider the mitigation measures proposed. Of course, it is sometimes difficult to limit ourselves to the ecological impacts because the intensification and expansion of this industry has many social, economic and ethical impacts that should be addressed simultaneously. The failure to combine the social, economic, ethical and environmental impacts produces an ineffective and contradictory set of solutions.

Greenhouse Gas Emissions

The livestock industry is responsible for generating between 4.6 and 7.1 billion tons of greenhouse gases each year, or between 15 and 24 per cent of total global GHG emissions measured as CO_2 equivalents (Fiala, 2008). This includes 9 per cent of anthropogenic CO_2 emissions, 37 per cent of methane (with 23 times the global warming potential [GWP] of CO_2) and 65 per cent of anthropogenic nitrous oxide (with 296 times the GWP of CO_2) (Steinfeld, et al., 2006). Livestock accounts for some 64 per cent of global anthropogenic emissions of ammonia, and processing may be a significant source of high GWP gases (such as hydrofluorocarbons) as well as of CO_2. GHG emissions are most pronounced from deforestation, enteric fermentation and manure. Cattle production accounts for most of the deforestation and much of the enteric fermentation. Pig and cattle production accounts for most of the methane produced from manure. These emissions are expected to grow rapidly as demand for meat and dairy products doubles over the next 50 years (Steinfeld, et al., 2006), and will continue to increase despite further intensification (Fiala, 2008). According to Subak (1999) and Nathan Fiala (2008), producing 1 kg of beef has a similar impact on the environment in terms of CO_2 as 6.2 gallons of petrol or driving 260 km in a mid-sized car.

Other Air Pollutants

Concentrated animal feed operations (CAFOs) are big producers of air pollutants. Major air polluting gases include hydrogen sulphide, ammonia, airborne particulate matter and volatile organic compounds (Hoff, et al., 2002; Sneeringer, 2009). Particulate matter includes faecal matter, skin cells, feed materials and the products of microbial action on faeces and feed materials (Thorne, 2002). These are linked to respiratory infections, infant respiratory distress syndrome, perinatal disorders and spontaneous abortion. Stacy Sneeringer (2009) found that a doubling of livestock numbers in an area was significantly correlated with a 7.4 per cent increase in infant mortality with damage to the foetus being the most likely promoter. In addition, acidification and nitrogen deposition arises from ammonia volatilisation in the soil after deposition, a large part of which derives from animal excreta. This can produce forest dieback and possibly other impacts although these are relatively understudied (Steinfeld, et al., 2006, p.83). Almost no consistent air pollution emission data exist for concentrated livestock production facilities, making ecological, public health and other implications virtually impossible to estimate or prevent (Thorne, 2002; Sneeringer, 2009).

Overall, Alexander Popp, Hermann Lotze-Campen and Benjamin Bodirsky (2010) have estimated that non-CO_2 GHG emissions will increase significantly until 2055 even if consumption is held constant at 1995 levels. While acknowledging the possibility of technological and production-based mitigation in reducing the level of such emissions, they conclude that this is not as effective as changes in dietary habits (Popp, Lotze-Campen and Bodirsky, 2010, p.451). Others have more optimistically argued that a combination of changes in manure management and manure utilisation, and changes in feed type can reduce the use of fossil energy and GHG emissions by 61 per cent and 49 per cent respectively (Nguyen, Hermansen and Mogensen, 2011, p.2561)

Water Use and Pollution

Global water use by the livestock industry is estimated at 16.2 km^3 with cattle using the largest quantity (11.4 km^3) (Steinfeld, et al., 2006, p.131). Drinking and servicing requirements represent only 0.6 per cent of all freshwater use, but when added through the food chain, estimated water requirements (still quite simplified to include only product processing [slaughterhouses and tanneries] and feed production) exceed 8 per cent of the global human water use. Feed production constitutes the largest portion: 7 per cent of total human use. Estimates of groundwater depletion from feed production total 15 per cent of global water depleted annually (Steinfeld, et al., 2006, p.167). Local depletion of groundwater aquifers from livestock production is prominent on the southern High Plains of the United States and in parts of India, China and Botswana (Brooks, et al., 2000). Seaboard Foods (the second-largest pork producers in the United States) was primarily responsible for a metre of decline in the non-renewable High Plains Aquifer in the Oklahoma Panhandle from 2001 to 2006 (Oklahoma Water Resources Board, 2007).

Water quality issues derive from both extensive and intensive livestock waste production. Animal waste can harm water quality through surface run-off, leaching into soils and groundwater, direct discharges and spills. The more intensive the production process, the worse the brew of chemicals discharged. Nutrients (nitrogen and phosphorus primarily), sediment (erosion), pesticides, antibiotics, heavy metals, chemical disinfectants and pharmaceuticals such as hormones are primary constituents of intensively produced animal facilities. Hormones, used to enhance growth, include testosterone, progesterone, oestradiol, zeranol, trenbolone and melengestrol (the latter three are synthetic). Recombinant bovine somatotropin (rBST) is the most prevalent hormone used to stimulate milk in cows. Endocrine disruption in humans is a hormone-based concern and aquatic systems are quite vulnerable (Raloff, 2002). Even under extensive production modes, streams and groundwater can be polluted by nitrogen from excreta. Land application of animal manure can lead to the accumulation of heavy metals phosphorus in soil. And dairy soil has been shown to be a reservoir of multidrug-resistant bacteria in the transmission of infectious disease from farm animals to humans (Burgos, Ellington and Varela, 2005). While metal accumulation in soil is context dependent, nickel, cobalt and chromium are detected in most samples of pig slurry (Suresh, et al., 2009). Also, pesticides used in feedstock production pose threats to soil and water quality, as do fertilisers.

Biodiversity

Livestock production affects biodiversity in number of ways, depending upon the mode of production and cultural context. Wholesale deforestation impacts on a vast array of plant and animal species. Pollution affects terrestrial and aquatic species. Some of the biggest fish kill catastrophes in American history have occurred from pig facility lagoon ruptures during large storms. For example, a billion fish were killed in a North Carolina river in 1991 and bulldozers were used to clean them off beaches. Some 240 km of Missouri's streams were polluted from pig CAFOs, causing 61 fish kills and killing more than half a million fish. Buffalo Lake National Wildlife Refuge in Texas experienced large fish kills on the refuge in the 1960s and 1970s due to field run-off and discharges from cattle feedlots. Eutrophication may be the biggest threat to bio-diversity, where cyanobacteria blooms and outbreaks of botulism and avian cholera also occur (McMurtry, et al., 1989). On the North Carolina Coastal Plain alone an estimated 124,000 metric tons of nitrogen and 29,000 metric tons of phosphorus are generated annually by livestock. Concentrations of hormones fed to animals may have great implications for health of aquatic organisms as research shows that even low-level exposure to select hormones can elicit deleterious effects in aquatic species (Kolpin, et al., 2002).

The global consolidation of pig genetics (see Chapter 3) has seen most factory farms producing Duroc, Landrace or Yorkshire breeds of pig. All over the world, genetic diversity has been cast aside in favour of the uniformity of the animal body which will 'react' efficiently with the technique and technologies of production.

Labour Issues and Workers' Health

CAFOs are difficult places to work. One pig creates as much waste as five humans so the gases, smells, bacterial particles and hazards of working with large unhappy animals are quite undesirable. Similarly, poultry houses with thousands if not hundreds of thousands of birds are notorious for ammonia pollution and dust hazards.

Based on a pilot survey of CAFOs in two Indiana counties in 2008, the ratios of animals to workers were approximately 95 to 1 in dairies and 3000 to 1 in pig operations; hourly wages averaged $12.27 in dairy, $15.54 in pigs and $12.94 in beef (Keeney, 2008). Average wages for all CAFOs surveyed were $13.88 per hour. Over two-thirds of the hired labour on the CAFOs surveyed were legal and illegal immigrants. Nearly 30 per cent had been hired within the previous two years to replace those who were terminated (60 per cent) and those who left for other opportunities or because they did not want to work in CAFOs. Because of the conditions in the factory farms and the low pay, owners or their managers often have to recruit workers from outside the communities in which they locate. Over half the workers in Danish pig factories (producing 27 million pigs in 2010) are from Eastern Europe (primarily Ukraine and Romania), achieved through massive recruitment. William Kandel and Emilio A. Parrado (2005) have even correlated the growth of new Hispanic settlements with the establishment of meat-processing plants.

For example, Seaboard Foods recruited workers to its farms and plants from Mexico and Central America (Barlett and Steele, 2001). Workers also came from Laos and Vietnam to the plant in Guymon, Oklahoma and the

Kansas farms also owned by the company whose major shareholders are the Bresky family from Boston. In Guymon, the company built housing for the workers and deducted the costs of housing and the meals the employees get at work from their $300 a week salaries. In 2001 one of its managers pleaded guilty to three counts of felony cruelty to animals, discovered by way of an undercover film-maker working at the farm. The film showed employees beating pigs with a metal rod and slamming their heads into the concrete floor. A newer film, also made undercover, shows sows desperately chewing at the metal bars of their cages which are covered with urine and waste. Some are lying dead, some bloody. Piglets have their legs duct-taped to their bodies and others are suffering from infections and abscesses (Kretzer, 2012). Again, worker neglect and cruelty are the case. But Seaboard Corporation treats the workers with neglect and cruelty as well. In the few bits of information available about its treatment of workers it looks as though there is no holiday during the first year, and if someone is injured on the job the company does everything it can to avoid taking responsibility (see, for example, *Raul Dominguez v. Seaboard Farms*). Meanwhile, the chief executive officer, Steven J. Bresky, is fabulously wealthy, a billionaire whose compensation for 2011 was over $5 million, and financial advisers encourage investors to closely follow Seaboard Corporation stock. The Humane Society of the United States filed a complaint with the Securities and Exchange Commission in 2012 citing misleading of shareholders regarding the corporation's abuse of animals.

In fact, getting evidence of the treatment of both animals and workers within CAFOs and slaughterhouses generally happens because one of the humane societies or animal rights organisations has undercover workers who film the situation or because an anthropologist or sociologist takes a job in a slaughterhouse for a period of time to do ethnographic research for a degree or publication (Foer, 2009; Pachirat, 2011). These undercover operations are threatened by the new 'ag-gag' or anti-whistle blower bills being introduced in the United States, as well as by the Animal Enterprise Terrorism Act, a national law that prohibits crossing state lines with even the intention of economically threatening an animal research or animal food production facility. Ag-gag laws are beginning to appear in Australia as well. The industrial animal production coalitions and their politicians support these pushbacks against the realistic (as opposed to staged) transparency provided by undercover filming and investigation. The industry is fighting back hard against any campaign to reduce meat eating (like 'Meatless Monday' which is now global, see Chapter 6). Such is the case even when they are promoted by poor public schools, like those in Baltimore, Maryland, in order to enhance child health (Barclay, 2009). Concentrated animal feeding operations emit several toxic gases, vapours and particles: ammonia, hydrogen sulphide, carbon dioxide, malodorous vapours and particles contaminated with a wide range of microorganisms including endotoxins, fungi and allergens (Heederik, et al., 2007). There are over 150 pathogens found in animal manure including *Bacillus anthracis* which causes anthrax, *Escherichia coli* which causes colibacillosis and mastitis, and *Salmonella* which causes salmonellosis (Hribar, 2010).

In a poultry house, for example, the dust contains feed and faecal particles, feather barbules, skin debris, fungal fragments and spores, bacterial and bacterial fragments, viruses and particles of litter. This type of dust is typically known as organic dust, since it is derived from materials formed by living organisms. A mysterious foam is plaguing pig operations, trapping methane and causing barn explosions that kill pigs and workers. When you think about the amount of manure (filled with antibiotics, hormones and other chemicals) concentrated in one place, you can only wonder how anyone can work in such a place – never mind having to deal with the miserable animals themselves. Gigantic confined livestock and poultry operations produce half a billion tons of manure each year, three times as much as produced by all the people in the United States (Food and Water Watch, 2010). With no sewage treatment required and onsite storage of huge amounts of such waste, one can only imagine the difficulty of working in such places.

In general, workers experience bronchitis, respiratory distress, interstitial lung disease, chronic obstructive airways disease and toxic organic dust syndrome (Hribar, 2010). Up to 30 per cent of CAFO workers may experience these diseases (Horrigan, Lawrence and Walker, 2002). A primary cause is a constituent of bioaerosols; endotoxins are components of the outer membrane of Gram-negative bacteria which are more resistant to antibiotics than are Gram-positive bacteria. Endotoxins are parts of bacteria that remain toxic after the organism is dead. They can increase disease severity acting on their own, causing degraded lung function and inflammatory responses, or acting as natural adjuvants to augment asthma and atopic inflammation (Liebers, Raulf-Heimsoth and Brüning, 2008). Levels of up to 15,000 EU/m^3 may be found in animal confinement operations (Heederik, et al., 2007). Maximum recommended levels within pig confinement systems for human and pig exposure are 0.08 grams/m^3 and 0.15 g/m^3 respectively (Bowman, Mueller and Smith, 2000). A study from Iowa shows an increased prevalence of childhood asthma on farms with increasing numbers of pigs (Merchant, et al., 2005). Studies from Canada suggest that women are more likely than men to develop asthma from working in CAFOs (Heederick, et al., 2007). Antimicrobial resistance and endocrine disruptors are also important areas of concern.

Farmers and workers in contact with animals heavily dosed with antibiotics and therefore colonised by resistant bacteria are at much higher risk to methicillin-resistant *Staphylococcus aureus* (MRSA). MRSA is demonstrably present in pig farmers and workers in the Netherlands, Canada, the United States and France (Armand-Lefevre, Ruimy and Andremont, 2005; Smith, et al., 2009). A study undertaken among Iowa pig farm workers and pigs showed 70 per cent of the pigs and 64 per cent of workers at one company (with 60,000 pigs and 18 workers) with MRSA. Pigs and workers at a second company's facilities with 27,000 pigs and seven workers did not have MRSA although they were resistant to penicillin, oxacillin and tetracycline, and all were susceptible to trimethoprim-sulfamethoxazole, gentamicin, levofloxacin, moxifloxacin, linezolid, daptomycin, vancomycin and rifampin. Most of these strains or MRSA belong to type ST398 which was found first in France and is now found in much of Europe, China and the United States. MRSA of a slightly different sequence has been found in dairy cows, veal calves and humans in Britain, Denmark and the Netherlands.

Streptococcus suis is another growing risk for animal workers and an emerging zoonotic pathogen. Causing meningitis, endocarditis and other diseases in both species, and hearing loss in humans, it is a major problem in factory farming of pigs worldwide. In July 2005 the largest outbreak of human *Streptococcus suis* occurred in Sichuan province, China; 204 people were infected and 38 died. Two outbreaks were reported from Jiangsu province in 1998 and 1999. It is the primary cause of bacterial meningitis in Vietnam which has a large pig industry.

Sharing diseases with domesticated animals is not a new thing: humans and animals have traded tuberculosis, common influenza and measles for hundreds of years. We now know that pigs, humans and poultry share viruses and that large confinement systems are breeding grounds for new viruses. For example, the USDA's Swine Influenza Virus Surveillance Program tested 12,662 samples from 3,766 pig diagnostic laboratory submissions collected from 1 October 2010 to 31 July 2012. Over that time period, 1,488 case submissions were identified as positive for influenza A which infects people and animals. The high turnover of workers at CAFOs and the constant introduction of unexposed animals to replace those going to slaughter means there are multiple pathways to exposure from mutating, reassortant viruses. A strain of H1N1 virus with cases documented in 170 countries during 2009 is an example of the sort of global influenza that public health experts worry about. CAFO owners have no incentive to report influenzas, however, because they are not 'reportable illnesses' as per OIE mandate and economic compensation may be lost as was the case in Alberta, Canada when a herd was found infected with a novel strain of H1N1. CAFO workers are not monitored and may be reluctant to report symptoms because of fear of harassment or loss of employment (Schmidt, 2009). Although little research has been done on CAFO workers and influenza, Gregory C. Gray, et al. (2007) found that CAFO workers were 50 times more likely to have elevated H1N1 antibodies than non-exposed controls. Their spouses were 25 times more likely to harbour these antibodies.

Slaughterhouses, also known as meat-processing plants, are very dangerous and difficult places to work in. The speed of the line makes it nearly, if not totally, impossible to do a job that accomplishes what is necessary for worker health and safety, for animal welfare and for food safety. The speed of the line means that knives and other tools are not sharpened and cleaned as much as they should be, which makes a worker's job harder and makes the meat more unsanitary. Some parts of the plant are hot and humid, while processing rooms are often frigid. The technological set-up assumes one size fits all, but that is of course not the case and many workers are sore or injured because they have to reach up or bend over all shift long. Most workers have to stand for the entirety of their shifts and the work is tedious. Injury rates in the United States have purportedly improved over the past decade but are still higher than most manufacturing industries (Bureau of Labor Statistics, 2012/2013). Union activists argue that this reduction is actually due to the reformulation of injury categories by the Bureau of Labor Statistics such that typical injuries in meatpacking are no longer included. Many injuries go unreported as a survey of over 400 Nebraska meatpacking workers found that 62 per cent had been injured in the past year (Nebraska Appleseed, 2009). Workers who complain about their injuries may face dismissal as may those who identify meat safety and animal welfare infractions (Human Rights Watch, 2004; Nebraska Appleseed, 2009; Pachirat, 2011). Undocumented workers are especially intimidated by management and are easy targets for exploitation.

The speed of the line, reported by Timothy Pachirat's breathtaking account of a beef slaughterhouse in Omaha, Nebraska, is a cow death every 12 seconds. Cattle and cows are supposed to be stunned before their throats are slit, but sometimes it takes multiple attempts with the retractable bolt gun and not all of them are dead before the workers start cutting off the hide. The size of the animals, even as carcasses, makes the work dangerous and there is potential for slipping on the blood and fat. Poultry lines can run at 145 birds per minute and the US government wants to speed them up to 175 birds per minute and have the companies rather than USDA inspectors inspect the meat for food safety. In one Nebraska plant, near the home town of one of the authors, nearly 53,000 pigs are processed per week. The speed in the United States is limited only by a sanitation law, not occupational health or animal welfare policies. Poultry come into the slaughterhouses and are hung live by their feet in shackles in near total darkness in order to quieten the birds. But the birds are afraid, flapping their wings, pecking, throwing dust, urinating and defecating – all of which cause problems for the workers. Cutters may have to make up to five cuts every 15 seconds or 20,000 cuts a shift and repetitive motion injuries are common. Psychological injury is also common as slaughterhouses are places of blood, desperation and death (Dillard, 2008). In fact, slaughterhouse hosting communities are sites of relatively higher violent crime rates compared with those hosting other manufacturing industries according to a carefully controlled study undertaken in 581 'right to work' counties within the United States from 1994 to 2002 (Fitzgerald, Kalof and Dietz, 2009).

In 2012 the US meat and poultry industry processed 8.6 billion chickens, 33 million cattle, 250 million turkeys, 2.2 million sheep and lambs and 113 million pigs (US Department of Agriculture, 2014). Globally, more than 150 billion animals are slaughtered every year according to ADAPTT's 'Kill Counter'. Slaughtering and packing employs about 486,000 workers in the United States alone. Global numbers are not available. Wages have not kept up with inflation. According to the Bureau of Labor Statistics, the median wage in the United States was $23,380 in 2010. Most workers came on with no experience and with less than a completed high school education. Many of the workers are immigrants, a situation that can bring a number of social problems. This situation may be replicated in other countries as well; for example, in Australia most of the workers at slaughterhouses are of Macedonian and Asian descent.

On the bright side, a new generation of meatpacking plants has been developed in New Zealand. The plants are smaller, cleaner, slower and safer. Animals are washed carefully outside of the plant so that faecal matter does not accompany them inside. Employees help solve problems and write job descriptions. The staff turnover each year is 7 per cent rather than the 100 per cent at many US plants because employees are well paid and trained. The line speed is 50–80 per cent slower but processing efficiency can actually be higher than in big plants. The smaller plants are made economically viable by their design, greater worker input, lower requirement for animals to operate at full capacity, and by the withdrawal of government subsidies for agriculture in New Zealand (Bjerklie, 2007). They produce higher-quality meat and are suppliers to McDonald's.

While workers' welfare in general is still dismal, it is arguably easier to mitigate and improve than animals' welfare (see Chapter 5). But what about the myriad environmental impacts of factory farming? What can we do to moderate these impacts?

The Political Economy of Environmental Impact Assessment: Hoofprint Analysis

The food system is a major contributor to climate change, biodiversity loss, deforestation, desertification, water consumption and pollution, air pollution and energy depletion. Livestock production accounts for a large percentage of the impacts. The FAO's landmark study found that 18 per cent of GHG emissions in 2005 arose from livestock production – more than the global transportation sector (Steinfeld, et al., 2006). Reducing the consumption of meat and dairy products can significantly reduce GHG emissions and other environmental degradation. Despite the fact that livestock product consumption is increasing at a global level, the scientific findings about livestock producers' contribution to climate change present a tremendous opportunity to funda-mentally dismantle an industry that is found significantly problematic in terms of public health, animal cruelty, worker compensation and safety, and the environment.

Studies on the impacts of livestock production and consumption are widely available (Nijdam, Rood and Westhoek, 2012; Weis, 2013). The ecological hoofprint, as it has come to be known, comprises the ecological inputs and outputs required and produced by a farmed or ranched animal under a parti-cular mode of production. Animals, of course, are living beings and require sustenance. They eliminate waste, as do other animals (like humans), in mul-tiple phases – gas, liquid and solid. They use space, interact with ecological systems and reproduce. Humans make most (if not all) of the decisions about where these spaces are, what ecological systems they interact with, what they are fed, how they are medicated, how they reproduce, and where and how their wastes are managed.

Magnitudes of inputs and outputs per animal differ depending upon climate, production system, type of animal, length of life and reproductivity rate. Within a bodily ecology, food types can influence quantity and quality of bodily emissions. Big animals like cows produce the most impact – especially if forests are cut down to provide pasture. Hoofprints also depend upon the concentration of animals. Even though the notion of a 'hoofprint' might suggest just the spatial effect, that is not the meaning of the term at this time. Early studies of human footprints looked at the amount of land that would be required to provide food for a particular diet and the energy required for a certain size of home. Now, usage of the term 'footprint' or 'hoofprint' can mean any of a number of techniques for estimating inputs and outputs from animal production and consumption.

Multiple forms of estimating impacts exist. Some of the first work was done using net primary productivity measures and net energy analysis during the 1970s (Ayres, 1999). The first use of the term 'ecological footprint' was by Mathis Wackernagel and William Rees (1996). Their study converted all consumption into the land used for production and included as well the estimated amount of land necessary to sequester the greenhouse gases produced. Ecological footprint studies do acknowledge that impacts may be distal, but footprint methodologies do not track such distal impacts and rely on data that are insufficiently granular to differentiate different food product sourcing strategies (see, for example, Tam, 2011 on San Francisco).

Life-cycle assessment is increasingly recognised as the preferred method of measuring environmental impacts of products. One of the first efforts to undertake a life-cycle analysis which incorporated not only resources or energy but also waste was done in the 1970s by the US National Science Foundation's programme, Research on National Needs (Ayres, 1999). Now International Organisation for Standardisation standards (ISO 14040 and 104044) provide guidelines for conducting life-cycle analyses. Durk Nijdam, Trudy Rood and Henk Westhoek (2012) have reviewed 52 life-cycle assessment studies of meat, milk, seafood and other sources of protein. All of the studies quantified the emission of greenhouse gas; 17 examined eutrophication and 18 land use. Conventional production methods were often compared with alternative methods, such as free-range. Despite a number of methodological issues, the authors found that ruminant meat has the largest impact in terms of both greenhouse gas and land use. Vegetable protein has the smallest impacts (per kilogram of product). Only some seafood from aquaculture (like mussels) are as low in carbon emissions and land use. Ruminant meat from extensive systems and seafood from energy-intensive fisheries have the largest carbon footprints per kilogram of edible product. Lobster trawling is very carbon inefficient. Free-range meat has a higher carbon footprint according to some studies. Beyond carbon emissions and land use, there are other issues that are important, such as animal welfare, eutrophication, emissions of pesticides, use of hormones and depletion of resources. Many of these present problems are of a different scale from GHG emissions.

Life-cycle assessment studies present us with considerable insight on protein comparison. At base, they tell us what we have known for some time: vegetable protein is much less problematic than animal protein. The nuances of aquaculture and seafood in general, or middle-range proteins like poultry and pigs, are quite useful for decision makers who are serious about lowering GHG emissions and land use. More specific problems such as local-scale hormones versus biodiversity loss from comparable systems are still difficult to resolve. Extensive animal production is higher in some environmental impacts, lower in others and better for animal welfare. And some biomes are only suitable for producing extensive ruminants because crops cannot be grown for more intensive feeding.

Who Is Paying Attention to These Data?

Movements to reduce meat consumption, like the Meatless Monday campaign, are increasingly popular and focus on institutional shifts (see also Chapter 6). Not only are universities, hospitals, restaurants and other institutions re-examining their meat and dairy consumption levels, they are also evaluating and changing their sourcing of such products, emphasising local small farms over well-established industrial supply chains. The United Nations has promoted a shift from red meat and dairy to more vegetable-rich diets to reduce GHG emissions and the Intergovernmental Panel on Climate Change found the largest potential for reducing emissions lies with consumers. Nevertheless, these efforts often elicit a backlash from consumers who claim their consumer rights and sovereignty are being jeopardised by institutional decisions to reduce meat consumption (Emel and Hawkins, 2010). These individuals seem unaware that consumer sovereignty is already something of an illusion because consumption is always a collective endeavour. Meat eating is not an individual action alone, but is mediated through a historical and cultural politics. Food choices at the institutional level are already highly constricted, especially as they relate to the political and economic clout of the meat and dairy industries. Consumption choice is thus not based on knowledge or desire alone but is always a collective action dependent on choice-sets and information made available through a variety of institutional actors. This state of affairs is itself a reflection and outcome of the governmentality of the livestock complex.

Another group who are listening to these life-cycle assessment and hoofprint conclusions include industry analysts or industry-funded scientists looking for multiple ways to fix manure quantities and qualities, alter the biology of the animals, and otherwise change the existing technology such that industrial animal production can go on as per usual, growing and spreading. Tending to focus on GHG emissions and eutrophication, these researchers and the industry to whom they report work on changing feed, cloning new animals and manure treatment. Enviropig, for example, is the trademark for a genetically engineered line of pigs that convert phosphorus more efficiently than their unmodified Yorkshire siblings.

A third group who should be paying attention to life-cycle assessment and other studies is policymakers. But as noted by the authors of a recent research paper for Chatham House, 'despite the scale and trajectory of emissions from the livestock sector, it attracts remarkably little policy attention at either the international or national level' (Bailey, Froggatt and Wellesley, 2014, p.7). They argue that 'international finance for agricultural mitigation is limited' and that

> [n]egotiations under the United Nations Framework Convention on Climate Change (UNFCCC) have overlooked livestock. Efforts to establish a specific workstream on agriculture have failed and talks have instead focused on a framework for reducing emissions from deforestation and forest degradation and enhancing forest carbon stocks in developing countries (REDD+).

At the national level, rather than support livestock reductions, most countries subsidise livestock production. Only Bulgaria and France, of the 40 developed countries listed under Annex I of the UNFCCC, have established a reduction target for livestock-related GHG emissions. Only eight of the 55 developing countries with nationally appropriate mitigation actions mention the livestock sector. Fortunately, Brazil is one of them and has established a reduction target for livestock emissions as well (Bailey, Froggatt and Wellesley, 2014). In a detailed and comprehensive case study of New Zealand, Christopher Rosin and Mark H. Cooper (2015) signpost various complexities in mitigating GHG emissions in its livestock industry – complexities which would have resonance and learning points in other parts of the world. In pointing out the complicity of the government and the livestock industry, Rosin and Cooper (2015, p.324) explain that the efforts to affect GHG emissions policies in the livestock sector have been shaped into a broad policy that encourages 'the pursuit of efficiency gains rather than a reduction in the volume of emissions'. They also point out the political and practical challenges of enacting a unified GHG emission policy across the livestock industry due to the 'significant variation among livestock production systems'. Many of these complexities and obstacles are attributable to the resilience of the political ecology and governmentality of the livestock system in New Zealand. Even something as fundamental as 'where the responsibility for GHG reporting should be sited within the production chain' becomes an intractable political question, with the government attempting to manage and appease the multiple and competing interests of various actors in the chain (Rosin and Cooper, 2015, p.322). In practical terms, it is also at present 'not possible to differentiate the volume of emissions between two otherwise identical farms where one farm undertakes action to mitigate livestock emissions and the other farm does not' (Rosin and Cooper, 2015, p.320). Above all, the entire system is oriented towards requiring farmers to report GHG, and there is no compulsion for them to proactively take mitigating action. Indeed, under a cap and trade system GHG emissions could be economised.

To be sure, while GHG emission policies have had a much longer history in other sectors of the economy (particularly refrigeration and transport), the incorporation of GHG from the livestock industry is comparatively recent. It is hence not surprising that its implementation is still besieged by political, practical and even scientific problems. Nonetheless, given the all-powerful meat complex, often made even stronger by a favourable regulating regime, we would expect any policies that aim to dent the profit-making potential of the industry to be strongly resisted.

The Biopolitics of Creating the Model Animal: Ramifications and Mitigation

Besides ethical accounting like hoofprint analysis, in their commodification process towards rearing the model food animal producers innovate production systems in many ways. These include transforming from a grass-fed diet to a grain diet and the injection of questionable growth enhancers into animals. E.V. Elferink, Sanderine Nonhebel and H.C. Moll (2008) argue that grain-fed livestock represents a higher environmental impact than other forms of feed, including feeding the animals food residue from other food processing industries like sugar beet and potato. While their argument is sound, there is no reason to suggest that the formidable complex of the modern livestock industry will proactively pursue this course of action – not least because it is highly doubtful whether there will be enough food residue to feed all the animals.

Susan Subak (1999) has argued for a meat tax that will take into consideration the negative environmental externalities of factory farming as a key mitigation strategy. While the negative impacts relating to GHG emissions of livestock intensification are not immediately visible or materially felt, this is not the same when such negative impacts affect health and environmental well-being in general. The question we ask is whether the concerns for health and food safety are the key to transforming the global meat factory (see Chapter 6).

Antibiotics, Growth Enhancers and Feedback Effect

Sneeringer, et al. (2015) estimate that about 80 per cent of the antibiotics consumed in the United States in 2011 went towards sustaining the livestock industry. The US Centers for Disease Control and Prevention report that more than 23,000 Americans die each year because of antibiotic resistance caused by the consumption of meat (National Institutes of Health, 2013). The emergence of 'superbugs' as a result of meat consumption is a public health concern (Schneider and Sharma, 2014). That the US livestock industry has revelled in this mode of production is testament to its lobbying strengths and the deep connections it has with public institutions.

Food scandals as a result of unsafe production practices (including the indiscriminate use of drugs) arguably have immediate impacts in dampening consumer demand. On 10 May 2006 the *New Straits Times* in Malaysia, in a short article, reported that the Consumers' Association of Penang had warned consumers to refrain from eating beef, mutton, duck and pork as tests had shown that they were tainted with salbutamol, a banned growth enhancer. All samples were bought at fresh markets in Penang. The same report noted that in June 2004 'the Health Ministry acknowledged that local farmers were using beta-agonists to produce more marketable lean meat and that an estimated 70–84 per cent of beef and pork contained the drug'. The Consumers' Association of Penang questioned the government as to why 'a substance such as a beta-agonist, a drug listed under the Poisons Schedule, could be obtained so easily' (Idris, 2006, p.27). The report detailed the use of beta-agonist on all types of livestock reared in Malaysia (except chickens). In the months that followed, it was reported that the authorities stepped up tests on pork sold in fresh markets and were imposing fines of RM2,000 ($490) on pork sellers whose meat was found to be contaminated with beta-agonist. This led to a 'pork war' between various actors and institutions in the pig sector. The chairman of the Malaysian Pork Sellers' Association (representing the individual pork sellers), Goh Chui Lai, was reported as saying that 'the government is punishing the wrong people. We buy pork from the farmers who are the ones who feed the animals. But why are we being punished when we are just selling the meat?' On the other hand, the secretary of the Federation of Livestock Farmers of Malaysia (representing the farmers), Sim Ah Hock, said no member farmer had been charged for using beta-agonist over the previous two years ago and that the farmers were innocent; he added that 'the Veterinary Services Department checks on pig farms and if farmers are found using beta-agonist, action should have been taken against them already' (*New Straits Times*, 27 October 2006, p.16). The crux of the matter appears to be that a minority of pig sellers are buying meat from a minority of pig farmers who actually used the banned, albeit still readily available, beta-agonist.

The fact that there have been no follow-ups on the use of beta-agonist in beef, duck or mutton, and that eventually only a negligible number of pig farms were found to have used beta-agonist, makes it appear that the action taken against the pig farmers was intentional and intimidating and a direct outcome of broader cultural politics (see Chapter 2). The protracted saga ended when all owners of the 656 pig farms in Malaysia agreed to sign a 'declaration' with the Ministry of Agriculture in February 2007 that pledged them against using any banned growth hormones on their pigs (Hamidah, 2007, p.20).

Food scandals such as the one discussed above can dampen the demand for meat, albeit often temporarily and in a highly localised manner. Indeed, the taste for meat is so entrenched that sporadic food scares are never sufficient for consumers to give up meat completely. Not least, with each successive scare, the authorities and industry actors invariably promise a round of yet more stringent regulation and more sophisticated scientific surveillance against future outbreaks (Andrews, 2013). While this is the typical way in which the industrial meat complex contains crisis, sometimes a debilitating virus attack can irrecoverably stunt the growth of the livestock industry. We draw again on the example of the Malaysian pig industry, specifically the Nipah virus attack it suffered in the late 1990s (see Chapter 2). As noted, the attack left a long-lasting effect on the industry which continues to be felt today. The effects lingered due to the overall political climate of the country which is ambivalent (to put it mildly) to the continuing development and expansion of the pig industry (Neo, 2009). This state of affairs is unique and reflects the specific context of the country. In any case, the setback suffered by the Malaysian pig industry made no difference, at a global scale, to the ever-increasing production and consumption of pork. Moreover, the frequent outbreaks of diseases in the livestock industry in general (Allen and Lavau, 2014) suggest that the creation of a model food animal through science and technology that is resistant to diseases remains a goal unfulfilled. Despite this, the architecture of the global industrial meat complex remains intact and indeed continues to flourish not-withstanding perennial food scares and zoonotic disease outbreak. In Chapter 6 we discuss other ways to potentially disrupt the architecture of global industrial meat complex.

Conclusion

Despite the clear ramifications of intensified production of food animals, the immense global market to feed the ever-growing appetite for meat has seen these problems only addressed in part and with no apparent urgency. As detailed in this chapter, an important reason is that many of these negative impacts are externalised to the environment. The strength of the global industrial meat complex and its complicity with national governments (some of the latter genuinely believed intensification as a strategy to increase the wealth and fulfil the appetites of their citizens) have seen efforts to redress such impacts as largely piecemeal, if not entirely ineffectual. The most recent attempt to 'de-externalise' the negative environmental impacts of livestock production makes use of the principle of ethical accounting to bring down the level of GHG emissions.

In any case, mitigation efforts such as hoofprint calculation (which in itself is a colossal policy challenge) and greater control over health and food safety have no *direct* relationship with moderating the demand for meat in the long run. In the next two chapters, we discuss how non-governmental organisa-tions and farmers who have opted out of (or resisted) the global industrial meat complex attempted to construct different futures for the production and consumption of meat.

5   The Thanatopolitics of Industrialised Animal Life and Death

Introduction

Thanatopolitics, most recently associated with the work of the Italian philosopher Roberto Esposito (2008), is the politics of the deaths of some in order to sustain the lives of others. Primarily focused on the Nazi idea of perfection and wide distribution of biotechnical killing methods, and the division of people by 'race' (broadly interpreted), this conceptualisation may also be applied to the modernist politics and techniques practised by many humans to kill animals such that they, the humans, may live better. These modernist practices of killing animals first began in the United States during the 1920s with the first indoor poultry houses hosting large numbers of chickens. Confining pigs followed in the 1950s and 1960s, at about the same time that big feedlots arose. Since then they have spread all over the world and are particularly prominent among new operations in Eastern Europe (see Chapter 2), Brazil, China, Vietnam, Thailand and parts of Africa. The FAO predicts that 80 per cent of the growth in global livestock production will be in the form of these intensified systems. CAFOs supply a steady stream of animals to industrial slaughterhouses, some of which can slaughter and dismember up to 20,000 animals per day.

Yet just as there were resisters of Nazi practices, CAFOs and industrial slaughterhouses are increasingly suspect in the eyes of many people – particularly on the grounds of animal welfare, worker exploitation and environmental implications (see Chapter 4). Graphic depictions of industrial animal spaces and practices are offered by a number of popular books and articles such as Jonathan Safran Foer's *Eating animals* (2009), Eric Schlosser's *Fast food nation* (2001), Gail A. Eisnitz's *Slaughterhouse* (1997), Timothy Pachirat's *Every twelve seconds* (2011), Karen Davis's *Prisoned chickens, poisoned eggs* (2009) and Ryan Gunderson's 'From cattle to capital' (2011), to name just a few. Multiple undercover investigations involving filming have been posted to the press or the web for viewing both slaughterhouse and CAFO conditions. At least one such film was submitted to the *Washington Post* by workers in a slaughterhouse (Warrick, 2001). While issues pertaining to animal welfare have been raised from time to time in previous chapters, here we specifically and more deeply consider animal welfare in the industrial meat complex commodity chain, from their rearing in CAFOs to their transportation to slaughterhouses (both distant and local). We especially describe in detail the happenings in CAFOs because these images and knowledges are often hidden and obscured by powerful actors in the meat complex. They do so to ensure that consumers remain oblivious of the insidious infractions of animal welfare in the production of meat, in the hope that values of empathy towards food animals will be lost on the consumers.

What Are CAFOs?

Concentrated (or confined) animal feeding operations (CAFOs), also called intensive feeding operations in Canada, are confinement systems where thousands of animals are housed and fed, such that as much of their metabolism as possible goes into making flesh, eggs and milk. It is the physical architecture that makes possible the extreme extraction of commodified animal bodies that we outlined in Chapter 3. The confined spaces purportedly allow for intensive management although many operations are almost fully automated and employ few workers – thus the individual is ignored and 'herd' or 'flock' sufficiency is the focus. The housing purportedly protects the animals from the weather, allowing for decreased bodily energy flow into keeping warm or cool. They are also protected from predators and some kinds of disease, although, as we have seen, other kinds may proliferate from the high densities and concentration of wastes. Giant fans and temperature modulation are used to keep temperatures amenable to 'efficient productivity' as well as to blow out the large quantities of effluent gases produced within housing systems. Waste disposal systems mechanically or hydraulically move the liquid and solid wastes from underneath the floors (or cages) to lagoons and other treatment facilities if such exist. Feedlots, where cattle are fed prior to slaughter, also concentrate large numbers to fatten quickly on grain and produce large amounts of manure waste and air pollution as well. Goats, buffalo and sheep may also be finished in feedlots. Poultry and pigs and even dairy cows may live their entire lives indoors without access to pasture, sunlight or space to roam. Antibiotics, vitamin D, insecticides and other chemicals are used to raise animals in these conditions, doing the job of sunlight and exercise (Noske, 1997). Workers, usually low paid and unskilled, are subject to noxious gases that produce multiple health issues, loud noise, and thousands of unhappy and possibly aggressive animals (Mitloehner and Schenker, 2007).

The litany of practices and architectures involved in industrial animal production is relatively well known. Dairy cows, for example, are artificially inseminated as often as possible, forced to stand on concrete or slatted flooring most of their short lives (and likely tethered or chained), bred to be very large and heavy, and to have large udders and high milk productivity vastly increasing the risk of painful mastitis. Tail amputation occurs in many locations, purportedly to eliminate uncleanliness although it reduces the capacity to get rid of flies and other insects. Denmark, Germany, Scotland, Sweden, England and some Australian states prohibit tail amputation (also called docking). The Canadian Veterinary Medical Association and the American Veterinary Medical Association oppose routine tail amputation in cattle, finding no benefit to the animal and no particular hygiene benefit. California, Ohio and Rhode Island have banned tail amputation in dairies but it is still common in several regions of the United States. Over 80 per cent of dairies in north-central and north-eastern United States were found to dock tails primarily for hygiene purposes (American Veterinary Medical Association, 2014). Dehorning is another painful practice that occurs in CAFO dairies to allow cows to feed closer together and purportedly to prevent skin lesions. Dehorning takes place with caustic paste, scoops, gougers, wires, hot irons and knives. Usually, as with tail amputation, no anaesthetic is provided. Because of their massive weight and the types of floors they typically stand on, lameness is a widespread problem (as it is with most CAFO animals). Another primary cause of lameness is ruminal (rumenal) acidosis due to grain eating rather than grass foraging, a problem that gravely affects feedlot cattle as well (Bergsten, 2003).

Poultry layers and broilers are confined in small cages (in the case of the former) or in huge houses with tens of thousands of birds mixed together in one room. The 1983 industry guidelines in the United States for caged laying hens stipulated 48 square inches (310 cm^2) per bird (the size of one piece of notebook paper). Even in 2007 the American Veterinary Medical Association suggested they would be satisfied with 67–86 square inches (432–554 cm^2) (Golab, 2007). Stocking densities can be up to 50 birds per square metre in poultry houses and this is for birds living possibly up to two years. These chickens are produced from incubators without ever seeing a parent or having an identifiable group. The lights are kept on more often than not to avoid breast blisters (from lying in ammonia-damp litter) and to encourage more feeding and drinking. Debeaked and claw-stunted laying hens are packed into small crates or similar kinds of 'open housing' under lighting scenarios designed to force maximum egg laying. Cages eliminate eggs laid on the floor, the need for litter and laying nests. They require less labour, basically less feed, and allow for increased bird density in a given area because they can be stacked. In the end, the birds may be ripped out of their cages leaving behind wings, heads and other body parts or collected by hand or mechanically and stuffed into crates for transport to the slaughterhouse where they will be killed by electrical water baths, carbon dioxide or carbon monoxide poisoning. Spent hens may not even be killed before they are composted as it is deemed too expensive to do so (Davis, 2009).

Pigs are separated into two groups – farrowing houses and pig-feeding operations. In farrowing operations, sows are impregnated as often as possible and produce as many piglets as possible. For the past few decades, in most places, they have been kept in gestation crates, narrow stalls that do not allow them to turn around. Temple Grandin, the slaughterhouse designer, likens living in a gestation crate to a lifelong sentence in a first-class airplane seat. 'You could maybe turn over on your side', she says, 'and there's someone bringing you food and water and everything you need, but you can't move. The pigs do not emerge unscathed, they can feel fear and pain'. When giving birth, they are moved to farrowing stalls which have an adjoining space for the piglets to stay for a few weeks without being inadvertently flattened by a sow so heavy she cannot completely control her body as she lies down. As soon as they are born, the piglets have their teeth filed or ground, their tails amputated and, if male, are castrated – all without anaesthetic. The teeth 'resection' (partial or total removal of a bodily organ) is so they do not hurt their mother's teats or each other fighting for the best place to suckle which is their natural habit (especially when litter numbers are so high). Their tails are amputated or docked because they are so bored and socially and bodily frustrated within their small pens that they bite each other. Imagine standing in a small pen day after day with nothing to do but eat on occasion. Tail amputation in free-range pigs is uncommon and unnecessary if they have enough room to get away from their more aggressive kin, have access to dirt for digging and have a stimulating environment (McCosker, 2012). In CAFOs none of these conditions is met, so tails are cut off with knives, cauterised or removed through ringing with an elastic band that kills the blood flow. The piglets, once they are weaned (and this happens quickly), move to pens where they spend the remainder of their few months being bored, frustrated and fattened.

For the typical consumer, these descriptions are far removed from the sanitised and commoditised version of meat that they purchase in the marketplace. This is not to say that with such knowledges, consumers will necessarily modify their dietary habits (see Chapter 6); rather, it is to suggest that the normalisation of meat consumption (especially 'cheap' meat) is predicated upon these hidden knowledges.

The Biopolitics of the Animal-Industrial Complex

Confinement, prevention of natural behaviours, mutilation of 'other' species of animals are biopolitical practices redolent of the thanatopolitics described by Esposito in *Bíos: biopolitics and philosophy* (2008). Esposito points out that Foucault illustrated how 'racism' was used to produce a separation between those who need to remain alive and those who are to be killed, because it is 'precisely the deaths of the latter that enable and authorise the survival of the former' (Esposito, 2008, p.110). In this monstrous thanatopolitics of the CAFOs, 'speciesism' rather than racism is the separation threshold. In fact, the extension of meat eating to all humans who desire meat is a way of practising a biopolitics of population control at the expense of the 'food animals'. One might argue that there is a regulationist role for meat as a stabiliser of class structures, just as there is for cheap oil (Huber, 2013). At any rate, considerable money (taxes and profits) goes into the research and technologies that support this vast enterprise of death for the sake of life. Arguments that animal protein is necessary for human health underwrite this animal-industrial complex enabled by capitalism. This is but one of the many arguments about meat consumption which activists have tried to debunk (see Chapter 6). These poor animals 'exist without life'. Esposito (2008, p.134) writes that 'existence for the sake of existence, simple existence is dead life or death that lives, a flesh without body'. This expectation of these 'food' animals, that they should exist without social ties, maternal bonds, space to move, the food their bodies have evolved to eat, exercise, renders them dead life – existence without life. Rachel Carson, in her foreword to Ruth Harrison's monumental book *Animal machines* (1964, p.viii) that first revealed the conditions within which these animals must live, asked:

> Has [man] the right, as in these examples, to reduce life to a bare existence that is scarcely life at all? Has he the further right to terminate these wretched lives by means that are wantonly cruel? My own answer is an unqualified no. It is my belief that man will never be at peace with his own kind until he has recognised the Schweitzerian ethic that embraces decent consideration for all living creatures – a true reverence for life.

The animals are victims of their own vitality – they go on living even when, according to any definition of 'life', they are dead. The sow's body continues to respond to artificial insemination or copulation and impregnation, time after time until she is no longer competitive with the young gilts who take their chance at producing enough piglets to satisfy the ever-increasing corporate goals. Despite her interminable boredom, her lameness, her illness, as long as she produces enough piglets, she goes on living in her cramped quarters. Slaughterhouse workers kill many female cows and goats that are pregnant. They may even be giving birth in the slaughter line while managers wait impatiently for the afterbirth to pass (Pachirat, 2011). Cancer researchers take the foetuses and their blood. Life is cheap because of its prolificity; culling and killing are nothing. Life is excess. Animals can be produced in the millions, especially birds which are not even considered animals in the United States according to the Humane Slaughter Act, the only federal legislation that protects farm animals. But life is more than metabolising food, reproducing, breathing, standing, lying down and dying. Life is more than flesh.

The Theory and Science of Animal 'Welfare'

So what is meant by animal welfare? The definition is a fiercely argued moving target. Generally, 'welfare' is understood to be reformist and not abolitionist when it comes to using non-human animals for food (Haynes, 2011). In other words, improving 'welfare' means continuing to raise and slaughter animals for food and other commodities, just doing it in a way that is 'better' for the animals. We will employ that difference in our discussion here and leave the abolitionist cause to the next chapter when we discuss vegetarianism and synthetic meat. Despite the reformist objectives, nevertheless, the economic stakes are high for specific conceptualisations of welfare, given that definitions used in supply chains and 'real food' indices segregate and stigmatise particular production practices (Bock and Buller, 2013). The suffering stakes are also high for the animals that must live in the conditions specified by particular definitions of welfare. In this section, we provide an overview of the science and policy of animal welfare as well as a short discussion of current thinking on human–animal relations that bears on food animals.

Animal welfare science is changing, reflecting and informing the evolving philosophies and politics of animal lives, as well as new methods of assessing animal capacities, feelings and desires. Essentially, three approaches to animal welfare are evinced by scientists, policymakers and activists. These are the productivist approach which is primarily focused on animal health or 'biological fitness' (Bock and Buller, 2013); the naturalist approach which focuses upon animals being allowed to practise many of their natural behaviours such as foraging and nesting; and the affective welfare approach which focuses upon the emotional state of the individual animal – attempting to minimise pain and suffering as well as promoting positive feelings (Fraser, 2009). In the first case, an emphasis on health and functioning has led to assessment methods based on rates of disease, injury, mortality and reproductive success. Stress-related measurements, both behavioural and physiological, have for decades been a primary way that animal welfare was ascertained (Hagen, et al., 2011). Measuring physiological and behavioural responses to mutilations, housing densities, handling styles and so forth became a major emphasis in science studies of welfare. These measures augmented those approaches that considered only productive and reproductive responses to the same kinds of actions – generally the most common animal industry approach (although body condition and injury are bottom line issues). In the second, an emphasis on affective states has led to assessment methods based on indicators of pain, fear, distress, frustration and similar experiences. An emphasis on natural living has led to research on the natural behaviour of animals and on the strength of animals' motivation to perform different elements of their behaviour (Hagen, et al., 2011, p.507).

In their review of animal welfare policies since the 1960s in Britain and the Netherlands, Bettina Bock and Henry Buller (2013) recount how a wider, deeper consideration of animal sentience and subjectivity has emerged. These authors write about the ontological shift accompanying and driving the 1965 issuance of the Brambell report, widely considered the first modern statement regarding farm animal welfare. The Brambell Committee was formed to look into the offences recounted in Ruth Harrison's book *Animal machines*, which was published in 1964 and provided one of the first glimpses into the horrendous conditions of the CAFOs or factory farms. The Farm Animal Welfare Council (FAWC), established in 1979 by the British government, used the findings of the Brambell Committee and other research and testimony, to devise a list of 'five freedoms'. The Brambell Committee's findings suggested animals should be able 'to stand up, lie down, turn around, groom themselves and stretch their limbs'. The FAWC extended these to include freedom from hunger and thirst; from discomfort; from pain, injury or disease; to express normal behaviour; and from distress (Farm Animal Welfare Council, 2009). The freedom to express normal behaviour is a significant move forward in the evolution of animal welfare. The Brambell Committee also set the stage for the science of animal welfare to 'take into account scientific evidence available concerning the feelings of animals that can be derived from their structure and functions and also from their behaviour' (Brambell, 1965). This science-oriented approach established the trajectory that 'welfare' studies were to take, with many cautioning against using an 'emotional' approach to welfare determinations. And, in fact, Bock and Buller stress that 'animal welfare science' is a classic case of the politicisation of science and the scientisation of politics with those arguing for a productionist approach to welfare often apologists for industry and those examining affect or subjectivity as too emotional – apologists for the animals perhaps?

Privileging any one of these freedoms has implications for confined animals, however, as confined animals of contemporary breeding may be ill adapted to 'perform' some of the welfare-promoting practices and accomplishing one type of welfare objective may exacerbate other welfare issues. For example, Marian Stamp Dawkins, Crystal A. Donnelly and Tracey A. Jones (2004) found that bird stocking densities in broiler houses did effect leg and footpad damage as well as jostling and growth, but that house age, ventilation, humidity, temperature and husbandry (number of 'stockman' visits) did as well. The authors cautioned against legislation that simply focuses on stocking densities without also considering 'environment' and husbandry or management. These researchers used bird mortality, gait, jostling index, footpad condition and faecal corticosteroid levels as indicators of 'welfare' pursuant to the different conditions mentioned above. Space augmentation for animals is a worthy objective, but it must be considered along with many other factors in CAFOs. A redesign or rethink of the entire enterprise is more in order.

The science of animal behaviour has greatly extended human knowledge of animal emotions and bonding. Ethology, a disciplinary area related to zoology, involves both laboratory and field research on animal social bonds, animal communication, animal culture and animal cognition. Although the field originated in the 1930s with the work of Konrad Lorenz, Nikolaas Tinbergen and Karl von Frisch, interest in farm animal behaviour and behavioural deprivation is quite recent – it has not been a part of the traditional veterinary and animal science education programmes. The International Society for Applied Ethology (ISAE) was formed in 1966 as the Society for Veterinary Ethology. The ISAE's journal, *Applied Animal Behaviour Science*, is considered the flagship publication of animal welfare science. The *Journal of Applied Animal Welfare Science*, sponsored by the Animals and Society Institute and the American Society for the Prevention of Cruelty to Animals, published its first issue in 1998. Posts have been created such as 'professor of animal welfare'. The Cambridge Declaration on Consciousness was put forward in 2012 by a prominent group of neuroscientists. The declaration concludes that

> non-human animals have the neuroanatomical, neurochemical, and neurophysiological substrates of conscious states along with the capacity to exhibit intentional behaviours. Consequently, the weight of evidence indicates that humans are not unique in possessing the neurological substrates that generate consciousness. Non-human animals, including all mammals and birds, and many other creatures, including octopuses, also possess these neurological substrates.

Many scientists and most pet owners, of course, have known this for some time. Studies have shown that chickens have empathy (Edgar, et al., 2011), and that birds in general are much more intelligent than common knowledge gives credit. Researchers at the University of Bristol have shown that chickens are born with the capacity to keep track of numbers up to five, that they can exercise self-control by turning down one reward if there is the prospect of receiving a better one, and that they have a preference for objects 'that they know make sense'. Others have shown that, unsurprisingly, chickens prefer clean air to that filled with ammonia. Research conducted at Cambridge University involved measuring cows' brainwaves with an electroencephalograph (EEG) machine while the animals had to find a way to open a door and get food. The test revealed that the cows became excited at the prospect of finding the food, suggesting that, like humans, they were stimulated by the problem-solving process. Vocalisation studies illustrate that cows and other animals know each other and other species, that they can be taught to come when called, and that individual calves recognise their own names. A famous pig study that was done in the 1980s with domesticated gilts and a sow ready to birth her second litter found the mothers-to-be wandered between 2.5 and 6.5 km over four to six hours looking for suitable nesting sites. They built nests within which they maintained their piglets for approximately nine days after which they rejoined the flocks. Their behaviours over the study period reflected those of wild boars and feral pigs (Jensen, 1986). Pigs like listening to music, like footballs, have complex social lives, can be taught the same tricks as dogs and enjoy getting massages (Crary, 2013).

   According to a review of the evolution of animal welfare science, it gained credibility slowly within the scientific community.

> From within veterinary science and animal science, animal welfare scientists seemed to criticise existing conditions and thereby carry the burden of not being independent and/or of being involved (too) emotionally. From within cognitive and neuroscience, applied researchers – in contrast

with laboratory researchers – were criticised for having less control over experimental conditions, and from within ethology, domesticated species were seen as poor research models.

(Hagen, et al., 2011. pp.94–5)

Nevertheless, ethologists, cognitive scientists and neuroscientists and others have proven that farm animals have needs, intelligence and emotions (positive and negative), and respond to rewards, punishments and a variety of other phenomena in definable ways.

Despite the challenging position of animal welfare science within the large academic and industrial communities, broad-based acceptance of the importance of farm animal welfare has emerged within the last decade as evidenced in the European research action Welfare Quality® project (Botreau, Veissier and Perny, 2009) and the European Parliament's Science and Technology Options Assessment annual reports (Bokma-Bakker, et al., 2009). Globally there is also movement towards acceptance. The Office International des Epizooties (which has since been renamed Organisation Mondiale de la Santé Animale [World Organisation for Animal Health]) convened a working group on animal welfare in 2002 and held its first international animal welfare conference in 2004. The FAO produced a report on capacity building for animal welfare (Fraser, 2009). Its goal is to help farmers in developing countries access markets that adhere to particular animal welfare standards.

'There is no word for "pig" in the Lakota language': Case Study of South Dakota

In this production unit, sited on land owned by the Sičháǧu Oyáte (Rosebud Sioux or Burnt Thighs Nation in Lakota) near the Rosebud Sioux Reservation, are 24 buildings housing upwards of 48,000 pigs (see Figure 5.1). Another facility close by is of equal size. The production facilities were built and operated by Bell Farms Group (also known as Sun Prairie). The original plan was to build 232 buildings on 445 ha to produce 800,000 pigs per year from 25,000 sows, using 6.4 million litres of water each day. The Rosebud tribal council – the elected body that governs the reservation – made a deal with Bell Farms, with a view towards income generation and 200 jobs in a place where generations of tribal members had not had the opportunity to work. Production began in 1999 in the first two facilities with about 18 employees, most of whom were Native American except for the management. Yet many tribal members resisted the Bell Farms development, particularly two outspoken women, Oleta Mednansky and Eva Iyotte, who were concerned about sacred sites and water contamination among other things. With the help of Robert Kennedy Jr and the Waterkeeper's Alliance, the Humane Farming Association and other organisations, the activists worked to get the Bell Farms Group off the tribal trust lands. Leaders like Neola Spotted Tail

Figure 5.1 A concentrated animal feed operation in South Dakota, USA. Note the
          opaqueness of the facility

challenged the decision of the tribal council to allow the farm to locate on
tribal land, saying that while it would provide jobs asking: 'What kind of job
is that for Lakota men? How long will they last? They won't be able to stand
the smell and they won't treat the animals that way'. Mednansky argued that
'confinement is not good for anyone, definitely not the animals, because they
don't understand. It's not their way of life, and it's not our way of life, either'.
This activism is a critical landmark for producing a discourse that unites
concerns for human and non-human animal well-being, particularly in
opposition to the predatory corporations that make profits from exploiting
people, ecologies and animals.

A former Bell Farms worker of two years (who is now a tribal council
member) recounted in 2001 what happened with manure spills and animal
welfare. He stated in an affidavit recorded for lawyers in Rapid City, South
Dakota, that he had seen the auto flush (which was supposed to flush wastes
out of the barn and into manure digesters) back up

> approximately five or six pens deep, which would leave the pigs' heads in
> the last – the last pen, just the heads are above the shit water. And there
> are a lot occasions where the pigs would be dead floating in the shit water
> and this after it's found. What happens is Todd Krogman, I seen him
> numerous occasions open the back – along with Trent Loos, open the
> back door and let all the shit water spill out onto the ground thus causing
> like a gully wash effect.

The man giving the testimony recounted that he had seen this happen in 11 of the 24 barns. The lawyer questioning the former worker asked about the workers. He responded:

> It's really offensive to work out there. Especially when you get these pigs which are dead and laying in there for weeks and stuff, and the smell is just unbearable. It – it causes you to puke. And I seen [sic] numerous employees out there puking on numerous different occasions.

He was told not to report the spills and he said 'for fear of my job, I didn't report anything … In fear of reprisal and when inspectors come in I was always scared to tell them for fear of my job'. According to the testimony, the time sheets of employees were being altered so that favourites got paid for more time and for time off.

Overcrowding of pigs in the buildings was another issue about which the former worker testified. There were (are) 24 buildings designed to hold 2,000 pigs each. But the former worker said he had seen over 3,000 pigs per barn on numerous occasions.

> I know they're overpopulated. I see how they suffer during the heat because of overcrowding because of no room to move. I see how a lot of them get trampled, get broken legs because of the overcrowding. I also am assuming and possibly for sure know that because of the over-populating it causes more shit to run down the pipe thus possibly backing these pipes up, which is happening a lot out there when I was working there.

The former worker had data on the numbers in the barns – anywhere up to 3,271 in one building. He said:

> And when you're going in there to work with these pigs and know how many is in there, it's hard, it's real hard, to work with that many pigs. There's no room to move. The consequences of overcrowding is pigs that die from heat exhaustion because there's no room for them to move or even to get to the water because they're laying in there like sardines. And when this happens a pig would usually get stepped on and get some injuries caused by a broken leg, a dislocated joint, and being just possibly smothered. I have seen on numerous occasions pigs panting just from the heat and I remember pulling out a couple pigs, you know, out of – I'd say an average of maybe six to eight pigs a day because of heat exhaustion and overcrowding.

He went on to say that the pigs pulled out would just lie in the aisle for a week to three weeks and they would start decomposing to the point where they explode and 'the guts flow out of the pig'. 'If they were humane, they would put it out of its misery. But on numerous occasions I seen the pigs just waste away. They just lay there for weeks until they die of either starvation or whatever'. The overcrowding was also reported to cause injuries to the workers because there is not enough room to move so there can be needle injuries or they can be knocked down.

A journalist who interviewed tribal members who had worked at the farm wrote in 2003:

> They describe a company that discriminates against Indians, paying them less than white workers and at times demanding they work extra hours for no pay. They say dust and ammonia emitted from fetid pools of manure that sometimes accumulate inside the barns made them cough until their ribs hurt and made breathing a struggle, even months after leaving their jobs.

Workers have also made videotapes of the conditions inside the barns showing animals so tightly packed in their pens that strong hogs begin to cannibalise the weak, eating off tails and ears. Other pigs are shown with soccer ball-sized abscesses hanging from their bellies. During some weeks, hundreds of pigs die, some employees say, from mistreatment and disease (Petersen, 2003).

The Humane Farming Association had the air inside the barns tested for ammonia. At their peak the levels were more than double the federal government's average limit for workers during an eight-hour shift. A maintenance worker complained about breathing problems.

> Anthony Barrera said he was so short of breath after quitting as a maintenance worker at the farm last year that for months he could not finish a sentence. A doctor, he said, prescribed a medicine inhaler usually used by asthmatics and told him that breathing the barn air could cause lung problems like those suffered by coal miners. When employees gather in the farm's break room, he said, there is a continuous chorus of deep coughs.
>
> (Petersen, 2003)

The employees say more than 50 piglets froze to death one night during the winter after they were left on a truck overnight.

> Mr. Barrera said the barns do not have enough heat lamps to keep the babies warm, so they jump on top of each other in a pile under each light. Sometimes, he said, those on the bottom smother, while those on the top burn to death.
>
> (Petersen, 2003)

Another worker recounted how sick piglets were killed by smacking their heads on the concrete floor. Pictures showed many pigs with abdominal ruptures.

The production facility manager denied almost everything, claiming that the ex-worker was disgruntled. He stated that overstocking pigs is a common practice and that their average death rate is around 3 per cent, which he claimed is low for the industry. The Humane Farming Association brought all of the testimony, pictures and video to the attention of South Dakota's attorney general. The Attorney General's Office put together a team of experts to investigate, but before they could do a surprise visit the Humane Farming Association made a national announcement that they had asked the state to look into the enterprise. When the inspection team visited, everything was cleaned up and only a couple of minor citations for water pollution were issued. After years of litigation, Bell Farms and Sun Prairie removed their pigs and leased the operation to ValAdCo, a company notorious for violating air pollution laws at its operations in Minnesota, having over 150 infractions in one month (Meersman, 2001).

What is the truth about the situation? It is nearly impossible to know for sure. The secrecy surrounding the operation was evidenced by the experience of a reporter and editor of *South Dakota Magazine* who stopped at one of the two facilities when Bell Farms operated it. He spoke with one of the white workers outside the office until security came. The white worker called one of the Bells who told her to tell the reporter to leave immediately and to confiscate his notes (Wilson, 2000). This type of security is typical of industrial production sites in the United States and there are laws in many states that prevent people from even taking pictures of the operations. Utah, Iowa, North Dakota, Montana and Kansas already have such laws.

New laws are proposed (and passed) to prohibit undercover videotaping of animal cruelty. Recent filming has shown hens caged alongside the rotting corpses of their sisters, workers cutting and breaking off chick beaks, workers in Wyoming punching and kicking pigs and flinging piglets into the air (Oppel, 2013). The authorities charged the operators and workers, with the egg company losing its contract with McDonald's. These ag-gag laws – another clear evidence of private–public complicity in sustaining the industrial-animal complex – would make it a crime to tape undercover or would require the tapes to be turned over immediately to local law enforcement which would make careful investigation impossible. Employees in some states would not be able to film either.

Animal Welfare in Slaughterhouses

In vivid and moving prose, which we quote at length, Ted Conover (2013, p.34–35) describes a typical slaughterhouse scene:

> It usually takes more than one try, as the animals duck down or try to peer over the side of the chute, whose width the knocker can actually control with a foot pedal. One cow, unlike the others, lifts her head up high in order to sniff the knocking gun. *What could this thing be?* It's her

last thought. The knocker waits until her wet nose goes down, then lowers the gun and *thunk*. She slumps, then gets hoisted aloft with the others. The knocked animals hang next to one another for a while, waiting for the chain to start moving – like gondolas at the base of a ski lift. From time to time an animal kicks violently, sporadically. 'They're not really dead yet,' says Carolina, which I can hear because she's close to my ear and it's slightly less loud in here. In most cases, apparently, what she says is true and intentional: the pumping of their hearts will help drain the blood from their bodies once their necks are sliced open, which will happen in the ensuing minutes. By the time the chain has made a turn or two, the kicking will stop.

Several workers have reported on the many animals that are scalded or bled or cut up prior to death or satisfactory stunning. In the United States, chickens and other poultry are not covered by the meagre sanctions of the Animal Welfare Act and may be starved and dehydrated prior to slaughter. They may be frozen in transport or succumb to heatstroke prior to entering the slaughter yard. Cattle are supposed to be ambulatory in the slaughter line and several undercover operations have shown that they may be tortured into standing with electrical prods and water hoses. A Chino, California plant video shows employees forcing water down a cow's throat to get it to stand, jamming cows in the eyes, kicking them, stabbing forklifts into them and other cruelties (Humane Society of the United States, 2008). In another undercover film of Central Valley Meat Company in California, cows are shown being suffocated, electrically shocked, shot in the head, dragged by one leg on a conveyer belt to slaughter and undergoing other abuses (Hauser, 2002). The company was shut down briefly and forbidden to send meat to the USDA school lunch programme. Most all of these abused animals were dairy cows too 'spent' to walk properly into the slaughter lines.

Two investigations have shown Tyson workers torturing chickens in multiple ways and urinating on the live shackle lines. In their defence, workers state that they are not allowed to leave the lines to relieve themselves and must at times do whatever it takes. One worker tells his story about how the birds come through at over 180 per minute and every fifth one is not stunned properly. This worker who was a 'killer' on the line – attempting to catch those birds not stunned and kill them before the scalding – recounts:

> There is blood everywhere, in the 3' x 3' x 20' [91 cm x 91 x 610 cm] trough beneath the machine, on your face, your neck, your arms, all down your apron. You are covered in it. Sometimes you have to wash off the clots of blood, without taking your eyes off the line lest one slip by, which they will ...
>
> You can't catch them all, but you try. Every time you miss one you 'hear' the awful squawk it's making when you see it flopping around in the scalder, beating itself against the sides. Damn, another 'redbird.' You

know that for every one you see suffer like this, there have been as many as 10 you didn't see. You just know it happens. You hope the machine doesn't break down or falter. You just want to get through the night and go home. But, it will be a long 2½ hours until break time. More than two hours of killing nonstop. At least a couple dozen chickens a minute at best. At worst, a whole lot more …

Many people who do this commit violent acts. They commit crimes. People who already are criminals tend to gravitate towards this job. You can't have a strong conscience and kill living creatures night after night.

(Butler, 2003)

Chickens, which are supposed to be stunned in an electrocution via saline water bath, are often conscious while they are being scalded to death. They are supposed to be shackled by the feet (although workers have been shown to throw them at the shackles which may catch their heads instead) and then dragged by machine through saline water (the salinity makes the current more penetrating) as they are electroshocked. But especially if they are not shackled correctly, they will not be stunned properly and may enter the scalding tanks still alive and bleeding (Dillard, 2008). Countless stories of workers torturing animals just for fun compel Jennifer Dillard to argue that workers are psychologically damaged by working at slaughterhouses because in the wider society cruelty to animals 'just for fun' is not acceptable.

In Britain a number of grocery chains have asked to have monitoring of the insides of slaughterhouses after Animal Aid made an undercover film that showed pigs being hit in the face with shackles, sheep being thrown into pens, pigs being burned with cigarettes, animals being shocked in the ear, mouth and anus, pigs falling from shackles into blood pits being dragged out and reshackled, sheep being decapitated while still alive, animals screaming and struggling to escape, ewes being suckled by lambs as they were stunned, and many other offences (Fowler, 2013). Animal Aid is calling for all slaughterhouses to have CCTV, and 118 members of the House of Commons signed an early day motion in January 2013 asking the British government to undertake such an initiative.

It appears there is a long way to go – at least in the United States where even as recently as 2010 the Government Accountability Office found that only 1 per cent of the Food Safety and Inspection Service (FSIS) budget goes to inspecting (and enforcing) the Humane Slaughter Act (US Government Accountability Office, 2010). The remainder goes to food safety, which is the other mandate of the same arm of the FDA that is supposed to oversee both in the slaughterhouses. Dean Wyatt, an FSIS supervisory public health veterinarian, told a House of Representatives committee that he saw stabbing of conscious live pigs while shackled, multiple electric shockings and beatings of downed calves and cattle, among many other violations. He shut down Bushway Packing in Vermont three times prior to the release of the now infamous undercover video showing day-old calves (from dairies) being tortured into the slaughter line. Inspectors interviewed by the Government Accountability Office for the 2010 report volunteered they needed more training on evaluating animal sensibility, what to do about sensible animals on the bleed rail, what to do about double stunning, how much beating is too much beating, how much electrical prodding is too much, what to do when electrical stunning fails, and how to deal with animals slipping and falling on the slaughterhouse floors. They also reported inconsistent approaches to evaluating 'egregious' beating and shocking.

The types of treatment found by animal rights and animal welfare groups within US slaughterhouses and CAFOs are present in other countries as well. Turkish slaughterhouses were found hoisting 270 to 360 kg cattle up by broken legs, dragging injured and sick cows out of trucks where they were electrically prodded but could not get up, and cutting the throats of thousands of animals with no stunning. The floors were so slippery with blood that large numbers of animals were slipping and falling (Eyes on Animals, 2013). Muslim and Jewish religious slaughter generally does not allow for stunning prior to 'bleeding', which can leave an animal alive for minutes with incredible pain (see Chapter 2). In the United States, the Humane Slaughter Act exempts religious slaughter, no matter what size the operation. In France, 80 per cent of sheep are slaughtered without prior stunning.

The chain speed is the culprit most authors and researchers point to – carefully avoiding identifying any of the workers as the problem. According to one ex-worker who began in 1979, the chain speed more than doubled in 20 years (Olsson, 2002). Profit is measured by chain speed and it is so fast that workers do not actually have time to worry about individual animals or about hygiene, although they are expected to in the United States, EU and elsewhere (Pachirat, 2011). According to accounts of people who have worked in slaughterhouses, the inspectors are anticipated and rarely examine the animals' situation, focusing nearly all their time on food safety issues. Unionising the workforce helps the workers get better wages and pensions but does nothing for the animals. In fact, a search of the United Food and Commercial Workers International Union website reveals not one mention of animal welfare. Union stewards can help the workers deal with issues on the floor; the animals need a steward of their own. In the United States, it will not be the USDA inspectors, who only expend 1 per cent of their efforts on human slaughter and the remainder on food safety.

Transnational Politics of Food Animal Transportation

Global trade in food animals is likely to increase as meat is produced in fewer places, albeit in larger quantities. This suggests an increased tension arising from the trade of live food animals, both in the way they are transported to importing countries as well as the way they are slaughtered in those countries. This is because, as evinced from earlier discussions, the moral ethics of food animals is variable across the world and animal welfare standards are likely to reflect such divergence in terms of how food animals are valued.

In May 2011 *Four Corners*, a popular Australian current affairs programme showed footage that revealed how precisely Indonesian abattoirs were processing cattle which were imported from Australia. The footage was filmed by Lyn White, the campaign director of Animals Australia, who visited 11 abattoirs in Indonesia in March 2011. The video depicts extensive scenes of animal cruelty, which contrasted sharply with a report commissioned by Ivan Caple, et al. (2010, p.44) that concluded animal welfare in Indonesia 'was generally noted to be good'. In Indonesia, out of the 100 abattoirs, only six employ stunning before slaughtering (Jones, 2011) and a majority of the abattoirs do not employ the method of stunning before slaughter given that Islamic law requires the animals to be conscious at the point of slaughter (Bruce, 2011). The traditional method of slaughter, which is to rope and drag the animal onto its side and on the floor, is condemned by the Australian Livestock Export Corporation (LiveCorp), a non-profit industry body that works in tandem with numerous other stakeholders to strive towards improving the welfare of animals (LiveCorp, 2013). Also, the traditional method is harder to apply to Australian cattle as they are bigger in size and hence much more challenging to control (Doyle, 2011). In 2000 LiveCorp worked together with Meat and Livestock Australia, a provider of marketing and R&D to the livestock industry, to design a metal restraining box for the purpose of slaughter without stunning (Meat and Livestock Australia, 2013). According to Jason Hatchett, a representative from LiveCorp, this design removes all risks involved in the restraining of the cattle and improves the efficiency and effectiveness of the process, which in turn makes it more profitable as well as reduces the stress in cattle. Funded by levies from cattle farmers and the federal government from 2004, 109 of such pieces of equipment were deployed across Indonesia, and Australian experts have been visiting the abattoirs to equip the workers with the necessary skills and knowledge to operate the equipment. Six visits were conducted in the 14 months after the filming in 2011.

However, the design of the restraint appears to have increased the risks and suffering faced by the animals. In an interview, Temple Grandin criticised the fundamental design of the restraint. With the slanted surface and wet flooring designed to trip the animal to fall to its side, she criticised it as inhumane and emphasised that the animals would be in much distress (Ferguson, 2011). According to the OIE standards, a non-slip floor should be implemented and restraints should not lead to sudden movements such as the tripping of the animal, thus rendering such design and practices unacceptable (Shimshony and Chaudry, 2005). The design of a trip box also did not fall within the acceptable methods of restraints presented by the OIE (World Organisation for Animal Health, 2014).

However, the inadequacy of the restraints was not the only source of concern in the video. The apparent pain and distress the cattle were subjected to before and during slaughter due to poor handling also appeared repeatedly throughout the film. This in turn suggests that the training conducted by the Australians is ineffective given that LiveCorp has had many years to improve the welfare of the animals ever since it started exporting livestock to Indonesia in 1993 (Doyle, 2011).

The video showed the inability of the workers to coax the cattle from the holding pens to the restraint boxes. The cattle were lashed with ropes, kicked and beaten up by the workers and some even had their tails pulled and were broken in the process. The distress felt by the cattle was clearly heard through their cries in response to such brutal treatment. In the Kota Binjai abattoir in Medan, one of the cattle bellowed as it fell repeatedly and was subjected to numerous instances of abuse. Another was roped according to the traditional methods and broke a leg as it slipped on the wet floor covered in faeces. For 25 minutes, instead of killing it on the spot, one of the workers tried all sorts of methods to try and get the steer to move, including breaking its tail, gouging its nose and eye socket despite its apparent state of collapse (Doyle, 2011). The steer was also kicked nine times and had water poured into its nostrils before slaughter. This clearly defies the OIE standards, which require any injured animals to be killed at the site where it is found (Shimshony and Chaudry, 2005). Such acts also go against OIE guidelines which state that 'under no circumstances should animal handlers resort to violent acts' such as 'breaking animals' tails, grasping animals' eyes' and applying 'irritant substance to sensitive areas such as eyes, mouth, ears' (Shimshony and Chaudry, 2005, p.697).

In numerous other cases, buckets of water were splashed on the fallen animals and most of the animals reacted with increased struggling as well as several head lifts. Resounding noises were also recorded by the video as the animals repeatedly banged their heads against the concrete floor. The average number of cuts to each animal was about 10, with the highest being 33 cuts. The reason for these serious injuries was attributed to lack of proper cutting skills as well as blunt knives, an evident failure to fulfil the Islamic injunction that the knife has to be sharp (Islam, 2007). As a result, the length of time taken for the animals to lose consciousness, which is also akin to the length of suffering after the cut has been made, can be as long as three minutes in Gondrong, one of the abattoirs filmed. All this is irrefutable proof that the welfare of the animals is dismal.

In one instance, for example, rather than abiding by the Islamic law of slitting the throat in one cut, the cut appears to be done in an extremely rough sawing action that resulted in the head being half-severed and the cattle remaining conscious and alive. The footage then showed the cattle sliding off the slaughter platform onto the floor and getting to its feet. With blood pouring from the gaping wound in its throat, the injured animal then charged towards the people. Animal cruelty was further evident when one of the workers slashed the tendon on the cattle's rear leg to prevent it from getting on its feet again, a form of restraint condemned by the OIE (Shimshony and

Chaudry, 2005). Bidda Jones, the chief scientist of the Royal Society for the Prevention of Cruelty of Animals, commented that there are indications of fear and distress as the animal was bellowing at that point and its tongue was hanging out (Doyle, 2011).

In another instance, in the Jalan Stasian abattoir in Medan, the animals were tied up and forced to witness the scene of slaughter of others. According to Grandin, animals do experience fear, and the observation made by Lyn White was that the last standing steer was trembling as it witnessed the scene. In all likelihood, these animals must have been able to sense what was happening around them and feel immense fear and trauma. More importantly, this goes against the Islamic requirement that no animal should be killed in front of another (Islam, 2007).

The video also questioned the possibility of improving animal welfare in the Indonesian abattoirs through the implementation of stunning. According to one of the workers interviewed, stunning is unacceptable because it is viewed to be akin to torture, according to Islamic law (Doyle, 2011). He also mentioned that the animals have to be laid down facing Mecca during the slaughter, a point that had not been raised in the literature covered. Even in instances when the abattoir owners agreed to implement stunning, protests by customers who deemed such processes as non-*halal* have led them to revert to religiously sanctioned slaughter.

The issues raised by the documentary footage are complex and should not be seen as a simple struggle between different ways of treating food animals as a result of different moral-religious and value imperatives. Not least, both Islamic *halal* and Jewish kosher methods of slaughtering are grounded in respect and the minimisation of pain to the animals, differing perspectives on the role of stunning notwithstanding. In most cases, the infringements seen in the video are rooted in economic expediency and clearly illustrate how cultural-religious values can easily fall by the wayside in the chase for more profits. The video exemplifies how contrasting expectations of the treatment of food animals will worsen with the increase in cross-border trade in live animals. It also highlights the difficulty of cross-border regulation and monitoring of best practices (assuming these exist and are agreed upon).

Be that as it may, the airing of the footage led to a wave of public outrage which forced the suspension of livestock exports to the abattoirs featured (Bryant, 2011). It was followed by a six-month suspension on shipments from Australia to Indonesia announced by the minister of agriculture, fisheries and forestry, Joe Ludwig. He reiterated that animal welfare remains one of Australia's top priorities and that the live food animal trade would only resume when it is ascertained that the slaughtering industry in Indonesia is able to comply with their standards. However, instead of the six-month suspension, trade resumed only one month after the ban, with new regulations being put in place to afford Australia better control over animal welfare in Indonesian markets.

While animal welfare organisations such as Animals Australia welcomed the ban, Indonesian officials argued that such footage unfairly created a negative image of Indonesia. Bayu Krisnamurthi, then vice-minister of agriculture of Indonesia remarked:

> I think ABC practised unfair journalism; only the bad practices were portrayed. We see this as an isolated incident. It's against the law in Indonesia to be so harsh and cruel to animals. Indonesian regulations explicitly say we must practise animal welfare in handling animals in the slaughterhouse.
>
> (BBC, 2011)

Indonesian officials also made claims of discrimination at the hands of the Australian government. They questioned why, despite many other countries carrying out similar practices, only Indonesia was hit with a ban (Iggulden, 2011). Indeed, officials in Australia also had differing views. Barnaby Joyce, a National Party spokesperson on regional development, claimed that the opposition coalition viewed such a move as an 'overreaction' that would punish some Indonesian abattoirs that had strived to comply with acceptable standards of animal welfare (Vasek, 2011). Meanwhile, an independent MP, Bob Katter from north Queensland, the heart of the beef industry, also criticised the Australian government for suspending trade, claiming that it was not focusing on the right issue. In other words, Australia should focus on demanding more humane treatment of animals in Indonesia's abattoirs instead of punishing Australian cattle exporters.

In 2009 Australia exported 80 per cent of its live cattle to Indonesia (Bruce, 2011) and about 60 per cent in 2010, with a value of A$319 million ($235 million) (Daley and Sedgman, 2011). Hence, a ban of livestock export to Indonesia can impact heavily on Australia's regional economy, a point noted by the minister of agriculture and food, Terry Redman. He shared similar concerns that the implementation of the ban was not a solution to the issue as Indonesia may import livestock from countries who 'don't care as much' as the Australians do (Daley and Sedgman, 2011, p.1), thus ultimately threatening the welfare of the animals.

Farmers in Australia also held differing opinions with regard to the ban. Some were worried that it would negatively affect their livelihoods (Bryant, 2011). The Northern Territory Cattlemen's Association was also concerned that a ban would only result in wastage of previous investments in terms of infrastructure and practices and would not necessarily address the issue of animal cruelty (Vasek, 2011). However, there were also those who viewed such measures as necessary for the government of Australia to protect the industry and sustain it in the long run (Bryant, 2011).

Beyond the slaughtering practices of receiving countries, the actual transportation of live animals in the global meat trade has also come under increasing scrutiny. In 2012 Bahrain rejected a shipment from Australia of approximately 22,000 sheep as some of the animals were infected with scabby mouth disease (*Doha News*, 2012). The sheep were stranded at sea for two weeks (Hernandez, 2012), threatening the welfare and health of the animals as the temperature reached 38°C. The decision made by Wellard, an Australian livestock export company, to unload the sheep in Pakistan without informing the authorities that the shipment had been rejected by Bahrain, proved to be both a blessing and a curse for the sheep (Ferguson and Masters, 2012). While the mortality rate of the sheep on board was ultimately well below the limit set by the Australian government, the decision ultimately led to the culling of all the remaining sheep under an order of the Pakistani authorities (Ockenden, 2012).

While the risk involved in the transportation process (in terms of the negative impact on animals) is generally known, this anecdote highlights another form of risk involved in the shipment of livestock to different destinations – the loss of control by the exporter who might be at the mercy of the importing authorities. In light of the Indonesian slaughterhouses and Bahrain/Pakistan import fiascos, more stringent measures had been put into place, such as the Exporter Supply Chain Assurance System that is supposed to guarantee Australian exporters control over their livestock right up to and including the slaughter (Ferguson and Masters, 2012). However, armed police forcefully removed Wellard and PK Livestock staff from an Indonesian abattoir that Wellard had invested in. Footage filmed by the public as well as PK Livestock workers revealed horrifying treatment during culling, including dragging sheep across the soil by their hind legs as well as live burials and sawing of the throats that left the animals half-alive (Ferguson and Masters, 2012). This contradicted the stand put forth in an interview with the secretary of the Livestock and Dairy Development Department in Pakistan who claimed that they ordered the culling to be carried out in the Islamic way. This incident also raised fundamental issues about the welfare implications of the transnational transportation of live animals across long distances. The journey from Australia to Bahrain for the live sheep is in excess of 9,650 km.

The Muslim dietary code contains prescriptions on animal slaughter which are geared towards alleviating an animal's suffering. Despite this, there has been relentless debate over whether religious slaughter is indeed humane (Shimshony and Chaudry, 2005). The level of pain inflicted determines if the animal is slaughtered in a humane manner. However, this simple and reasonable statement is deeply contested. First, and recalling our discussion of the science of animal welfare, what precisely is the level of pain that is acceptable in the slaughtering process? And second, how do we start measuring this pain? To some extent pain is fluid and subjective. There is nonetheless broad agreement that the humane slaughtering of animals means reducing the level of pain felt by an animal, before its death, to a minimum. For example, Neville Gregory and Frank Shaw (2000, p.216) argue that humane slaughter means that the animal loses total and permanent consciousness almost immediately after a fatal cut has been made. In that sense, the animal would have no time to feel any pain at all. Conversely, inhumane slaughter occurs 'when the animal is still conscious and has a functioning brain capable of perceiving pain or fear'. Extending this, it would seem that the most humane slaughter would be to render the animal unconscious – in a painless manner – before administering the fatal cut. However, for religious slaughtering in the kosher and *halal* traditions, the animal is required to be fully conscious while its throat is being slit before exsanguination (Bruce, 2011). According to Islamic guidelines, a prayer is also required to be said before the slaughter is carried out, in addition to various other requirements (Islam, 2007, pp.42–3) (see Box 5.1).

Box 5.1 Sample guidelines for Islamic slaughter

- The knife should be sharp-edged.
- The knife should not be sharpened in the sight of the animal.
- The animal should not be brought by pulling it forcibly; rather it should be brought with ease and should lie down comfortably.
- No animal should be slaughtered before another animal.
- The animal should be slaughtered immediately after making it lie on the ground.
- The animal should be left to cool down after slaughtering.
- The animal should not be slaughtered from the rear side of the neck.

Source: Islam, 2007.

For kosher slaughter similar guidelines aimed at reducing pain are also implemented, such as the requirement of a single cut severing both the jugular vein and carotid artery at once (Grandin and Regenstein, 1994). Arguably, religious slaughter fundamentally aims to minimise the suffering of the animal, treating each animal as more than an unfeeling commodity. Yet the knowledges about humane slaughter and such religious slaughter often contradict each other. We highlight the contentious issue of stunning to illustrate this point.

Religious slaughter often rejects stunning prior to slaughter. For example, Temple Grandin and Joe M. Regenstein (1994, p.119) argue that stunning prior to slaughter is unnecessary because religious slaughter without stunning can be humane if carried out with proper restraints. They claimed to have observed that animals 'had little or no reaction to the throat cut' and, aside from 'a slight flinch' from the initial encounter with the blade, it appears that 'the animal is not aware that its throat has been cut'. In contrast, others argue that stunning before the actual kill is clearly a more humane method of slaughter. Neville Gregory, Martin von Wenzlawowicz and Karen von Holleben (2009) find that animals that are not stunned or remain conscious after cutting are prone to suffering from airway irritation as a result of blood aspiration into the upper respiratory tract and lungs. Furthermore, Gregory and Shaw (2000) report that stunning by a penetrating captive bolt pistol when properly carried out has a high probability of leading to complete brain failure and thus almost instantaneous unconsciousness which will imply a minimisation of pain and suffering.

Some studies have also shown that developments in technology have enabled stunning processes to meet religious requirements. The development of sophisticated stunning equipment in New Zealand that adheres to Islamic requirements has led to the gradual acceptance of stunning in almost all sheep abattoirs in New Zealand and Australia (Grandin and Regenstein, 1994). However, the acceptance of these technologies remains limited and not at all widespread, due largely to differences in interpretation of religious imperatives and the very understandings of these technologies in the first place.

In sum, the existing literature shows inconsistencies in research about humane and religious slaughter and that what is considered humane is open to contestation. As noted, for humane slaughter, a key consideration is the length of time the animals take to lose consciousness after the cut (Gregory, von Wenzlawowicz and von Holleben, 2009). It is then assumed that during this time the animal is likely to be in pain or in distress. However, determining the time taken for animals to lose consciousness involves a variety of experiments. An electroencephalograph (EEG) has been commonly used to check for brain failure and electrocorticography (ECoG) to check for unanticipated activity. Based on these tests, studies have shown that calves experience brain failure immediately, suggesting little or no suffering when slaughtering them (Nangeroni and Kennett, 1963; Gregory and Wotton, 1984; Gregory, von Wenzlawowicz and von Holleben, 2009). Nonetheless, other livestock show longer and variable periods of time before they display no more signs of activity, as measured through EEG or ECoG (Newhook and Blackmore 1982; Blackmore 1984; Daly, Kallweit and Ellendorf, 1988; Bager, et al., 1992). Would these animals have suffered less had stunning prior to slaughtering been used?

While stunning before the slaughter is one of the ways that the pain and distress experienced by the animal can be minimised, there are disagreements in the treatment of pre-slaughter stunning within the religious communities. For some, stunning is acceptable and is an act of kindness and consideration, the caveat being that the stunned animal must still be alive before the slaughter. It is ironically this caveat that prevents others from endorsing stunning as they feel that the risk of killing the animal accidentally through stunning is ever present – thereby contravening the religious dictate that the animal must be alive prior to the administration of the fatal cut. Furthermore, it is also believed that stunning reduces the amount of blood loss and thus fails to adhere to the religious requirement that the animal needs to be completely bled out (Bergeaud-Blackler, 2007).

Studies have also emphasised that slaughterhouse workers can facilitate humane slaughter through proper handling of the equipment and animals. To ensure effective stunning, Gregory and Shaw (2000) also emphasise the need to train workers effectively so that they are equipped with the relevant technical expertise to carry out the process of stunning and are able to cope with unexpected situations, for example when repeated stunning is required. There must also be more stringent and consistent checks that the stunning is effective. The maintenance of technical equipment in this regard, as well as having back-up equipment at hand, is critical. These are calls that are echoed by Grandin and Regenstein (1994).

The foregoing discussion starkly illustrates how even for religious slaughtering – which is but a small component/technique in the entire industrial meat complex – both scientific and moral knowledges are politicised spatially. As a response to improving the welfare of food animals at the point of killing, humane slaughtering (as exemplified through religious slaughtering) is contentious. In fact, we have already sidestepped the equally important welfare implications of transporting food animals across a few thousand kilometres.

Conclusion

It is clear that there is a large toll to be paid for the pleasures of cheap meat, beyond the extensive environmental impacts discussed in the previous chapter. As we learn more about the capacities and social lives of animals, we see that welfare becomes a multidimensional panoply of needs, wants and physiological conditions. Yet, while the farm is the primary site in which species 'collide' and difficult questions over animal welfare and human morality need to be answered (Buller, 2013a), food animals range beyond the farms they are put into, as when they are transported to slaughterhouses, sometimes halfway across the world. We see that CAFOs and high-volume, fast-paced slaughterhouses are not good for any animals – human or non-human. And what regulations and guidelines that exist for welfare are not often met. Why else would the industry try so hard to keep people from seeing what really happens inside CAFOs and slaughterhouses? The growing global trade in live animals (which itself is rooted in a race towards the bottom of ever lower costs and the pursuit of higher profits) has also produced intractable problems in the welfare of animals when they are transported such great distances.

More importantly, besides enacting ever more welfare regulations which beg for much focused and sustained enforcement, free from the meddling arms of powerful industry interests, what else can we do to improve the welfare of food animals? Indeed, even when welfare standards can be universalised and enforced globally (which is a distant pipe dream), they do very little to dent the insatiable demand for meat. In the next chapter, we explore the possibility of not eating meat, organic livestock farming as well as the scientific possibility of producing synthetic meat. These are responses that aim to nip the problem in the bud by attempting to rid consumers of their taste for the meat of food animals.

6 On Not Eating Meat

Vegetarianism, Science and Advocacy

Introduction

In the previous chapters, we have detailed the negative repercussions of the industrial meat production system and how these consequences – especially the rendering of food animals as commodities – affect places, peoples, environments and animals' lives. Measures aimed at mitigating the array of negative impacts, for example, animal welfare regulation and environmental technologies (including environmental accounting and taxation) have proved to be limited and patchy in terms of success, particularly when assessed in their ability to fundamentally change and overcome the *normalisation* of industrial meat consumption. In this chapter, we look at two developments that aim to precisely and directly denormalise industrial meat consumption. The first posits organic meat production as a solution to the myriad of negative impacts of factory farming. Thus, it is offering an alternative process of producing meat, aimed at disrupting the homogenising *techniques* of production. The second strategy appeals to the value of meat avoidance/vegetarianism as a possible means to fundamentally destabilise and debunk the very notion of eating meat. We highlight two specific ways that this can be done, through advocacy and through substitution in the form of synthetic meat and meat analogues, both of which have to be made possible, ironically, by improvements in science and technology.

Using the case studies of vegetarianism advocacy and organic meat production in Asia, we suggest that the crushing governmentality of meat consumption and production works against the ultimate goal of reducing (industrial) meat consumption. For organic meat production, we show how the tendency for 'industrialising' organic farming is ever present – a testament to the resilience and inventiveness of the profit-seeking political economy of the livestock industry. This raises serious doubts about whether the welfare of the food animals will be secured for the long term. It also problematises the implicit assumption that organic meat production will necessarily lead to a significant reduction in meat consumption. For vegetarianism advocacy, we expose the difficulties in reforming entrenched tastes which relegate vegetarianism to its perennial marginal position. We then examine the links and contradictions between vegetarianism and technological interventions in nature – in the form of synthetic meat and meat analogues.

The Limits of Organic Livestock Production

While intensified production is increasingly the norm in the livestock industry, it is highly variegated in both scale and space. In most places, alternative forms of small-scale livestock production coexist with industrial ones (see Colombino and Giaccaria, 2015, for an example of the production of specialised Italian cattle). In yet other places, the (industrial) livestock revolution has yet to gain a meaningful foothold in the food economy. As Ian MacLachlan (2015, pp.22–3) observes: 'The most striking geographical feature of the livestock revolution is that it is not a global phenomenon so much as it is a regionally specific outcome of massive change in livestock production in a handful of extremely large developing countries.' Yet, because of the massive pent-up demand and 'growth potential' of these countries for meat, the resultant impact is still devastating, as we have tried to show in the preceding chapters. In this section, we again turn our analysis to China. China is the exemplary country to understand rapidly transforming meat production systems and the insatiable appetite for meat (Schneider and Sharma, 2014). Hence the case study of organic pig farming in south-west China is particularly useful to illustrate how even in a country where the state apparatus at the highest scale explicitly favours industrial meat, there *might* exist a space for niche meat production.

Organic livestock farming is a common non-abolitionist response towards factory farming and is aimed specifically at increasing the welfare of food animals in the rearing process. Organic farming is also believed to have the added advantage of reconnecting the links between rural agricultural landscapes, local communities and animals which have been severed by the dominance of large industrial farms (Dickson-Hoyle and Reenberg, 2009). To be sure, non-intensive small-scale commercial livestock production has always had an important role in the rural developmental process. Not least, the intensification of meat relies on a relatively high level of technical and infrastructural competence that is not yet readily available in many less developed regions, especially mountainous areas. Key issues that emerge in the literature include the viability and acceptability of small-scale livestock farming in rural communities (Mann and Kogl, 2003); the importance of small-scale livestock farming in alleviating environmental and food safety concerns (Hall, Ehui and Delgado, 2004); and the production of and preference for 'local' breeds of animals in small-scale organic livestock farming (Yarwood and Evans, 2006). Interestingly, the role of local government in organic livestock development of varying scales has often been viewed uncertainly, as either enabling or restrictive. In the discussion that follows, we point to an ambivalent governmentality that seemingly promotes small-scale organic livestock farming as a form of social-cultural practice (which can then be used as a poverty alleviation measure) but worryingly and ironically demonstrates considerable potential to undermine such efforts at the same time. The research is based on primary fieldwork done in south-west China between 2007 and 2011, and the discussion is drawn in part from Neo and Chen (2009).

China is the world's biggest consumer of pork. In 2012 its population consumed 50.4 million tonnes of pork, five times more than Americans, and China alone slaughtered 672.3 million pigs in 2011 (Compassion in World Farming, 2013). To be sure, in terms of overall consumption of meat China is ranked fifteenth in the world with per capita consumption of meat in 2013 standing at 48.8 kg (Myers, 2015). In comparison, Australians consumed the most meat at 93 kg per capita. In other words, the demand for meat in China is driven almost exclusively by its taste for pork. As noted in Chapter 4, compared with other major pig-producing countries, China's pig industry is relatively underdeveloped, although rapid modernisation of pig production is imminent (Schneider, 2011). Nonetheless, many rural households continue to rear pigs for subsistence or modest supplement to their household incomes. In many cases, these breeds are indigenous and not commercial.

Many indigenous breeds of pigs remain well known regionally, if not nationally. The culture of consuming local breeds is not unknown in China. These include the Jinhua pig, famed for its cured meat (Jinhua ham), the Taihu pig and the Wenchang pig. The latter, along with the South Yunnan small-ear pig (known variously as the Diannan small-ear pig or small winter melon pig), is found in south-west China. The small-ear pig (the focus of our case study), in particular, is highly confined in its distribution, found only in southern Yunnan province. Given its rich biodiversity, China has long recognised the importance of protecting native breeds. While appreciating the fact that 'a well-planned and systematic crossbreeding program may be essential for exploiting hybrid vigor to increase productivity', it also recognises that inadequate or uncontrolled crossings with exotic breeds would negatively impact on the quality of the meat. Worse, such experimentation might ruin the genetic integrity of the native breeds irreversibly (Zheng, 1985, p.210).

Thus, among other measures, indigenous breeds have been accorded protected status to ensure that their stock remains unadulterated. For example, in 1993 the small-ear pig was declared a national level two protected species. On the whole, native pig breeds in China share some common features: they are highly adaptable to harsh environments and are able to consume roughage. However, with most breeds having average litter sizes of eight, the small-ear pigs are generally low in productivity (Zheng, 1985, p.165). By various counts, such food animals have attained a certain level of cultural status and significance. Organic farming in its focus on niche breeds can also be seen as a response to the homogenising genetics of food animals (see Chapter 3). Nonetheless, as early as the 1920s agriculturalists sought to improve local breeds with superior Western ones (Schmalzer, 2002), although presently 'there is a lack of data about the percentage of the total market taken up by "western" or pink-pig pork – as opposed to the amount retained by the 40 or so popular native pigmeat breeds' (*Pig International*, 2007, p.19). Others have estimated that from more than 100 indigenous pig breeds available for sale in the 1970s, the Chinese pig market is now dominated by three imported breeds – Duroc, Landrace and Yorkshire (Schneider, 2011, p.9).

Clearly, the still slow but eventually inevitable 'structural shift to larger and more efficient units catering to the demand from the country's towns and cities has *tended to push smallholders aside*' (*Pig International*, 2005, p.11, emphasis added). We acknowledge such a tendency but nonetheless argue that, given the right context, there remains space for smallholders to survive in the imminent livestock revolution in China, especially if the latter are able to draw on values-based sociocultural imperatives. Not unlike the developments in wealthier nations, niche production of indigenous breeds of pig can meet the more discernible taste of a growing, albeit still limited, Chinese middle class in a segmented meat market. Moreover, such a development can potentially alleviate the poverty of rural villages, while at the same time extracting little pressure on both household resources and the environment. It hence fulfils the broader goals of social-political stability of otherwise troubled areas. It is with this latter hope in mind that local authorities have encouraged marginalised minorities to partake in niche organic livestock farming. This is an initiative that appears to dovetail well with the broader developmental strategy of poverty alleviation in the highlands of China, which attempts to connect rural spaces to urban centres and one actively encouraged by global developmental institutions like the World Bank (Neo and Pow, 2015). Moreover, as described earlier, backyard farming is also not an unfamiliar concept to the Chinese.

In the ensuing discussion, drawing on a case study of rural south-west China, we show the promise and limits of organic farming in disrupting the architecture of industrial meat complex in China. In the summers of 2007 and 2011, as well as February 2008, we visited Jinghong city and Jinuo Mountain (see Figure 6.1). Jinghong is the largest city in Xishuangbanna and about 540 km from the provincial capital, Kunming. Through personal contacts, we spoke to officials of the Agricultural Department of Jinghong to understand the status of the small-ear pig-rearing programme and also took several trips up Jinuo Mountain to meet informally with villagers who are participating in it.

The Jinghong Agricultural Department introduced the programme to the Jinuo people (one of the smallest ethnic minority groups in China) for several reasons. First, it believes that rearing small-ear pigs will not demand too many household resources from the villagers, whose main economic activity hitherto has been tea and rubber plantation labour. Second, it calculates that there is an untapped market in this indigenous breed of pig where a strong social and cultural demand exists for it. Last but not least, because of the proximity of Jinuo Mountain to the city itself, it envisages that transportation of the pigs would be manageable. For the officials, this scheme was work in progress.

The Jinghong Agricultural Department itself has a small-ear pig holding farm located 9 km from the city (see Figures 6.2 and 6.3). The farm was established in 1980 with the explicit aim of protecting the genetic purity of the species. Presently, the farm has an average of more than 500 pigs (including

Figure 6.1 Jinghong and Jinuo Mountain, Xishuangbanna, China

over 100 sows). The Agricultural Department also owns a non-operational farm in the neighbouring town of Menghai. The 10,000-pig-capacity farm was built in the late 1980s and was meant to rear conventional pigs.

In 2006, recognising a demand from well-off city residents for organic (*wugonghai*) meat, the Agricultural Department established a specialised butcher's shop in Jinghong city to sell the meat of the small-ear pigs (see Figure 6.4). To ensure a ready supply of pigs, and as part of a poverty alleviation measure by the city authorities, the Agricultural Department encouraged the Jinuo people to rear the pigs from fallow to finish. The pigs are then sold to the Agricultural Department for sale at the city butcher's. The department not only dispenses advice on how to rear the pigs but also provides each interested household a pair of sows and several piglets free of charge. (Incidentally, this mirrors the early years of the development of the Malaysian pig industry, elaborated in Chapter 2.)

Figure 6.2 Entrance of the holding farm for small-ear pigs in Jinghong, China

Figure 6.3 Tagged pigs in the holding farm for small-ear pigs in Jinghong, China

Figure 6.4 Specialist butcher's shop in Jinghong, China. The two columns of words in green read: 'Safe meat, organically raised; strengthen your spirit, return to nature.'

The village of Jiama saw 26 out of the 27 households involved in the small-ear pig-rearing scheme. This rate is by far the highest among the eight villages that participated. The high adoption rate is mostly due to the personal relationship that the village head has with the Jinghong Agricultural Department through which the former has previously supplied small-ear pigs to one of the departmental officials on an ad hoc basis. Most of the households have about four to six live pigs each with several having up to 10. When asked what attracted him to rear the small-ear pig, a villager from Jiama said:

> The pigs are very easy to rear and we don't have to put in much time or money to rear them. We give them [commercial] feed for the first two months and that costs some money but after that for the next five or six months, we just give them banana trunks which we mashed into pulp [see Figure 6.5]. When they are ready, someone [from the Agriculture Department] will come collect them. So it's not complicated.

Clearly, producing the small-ear pig is dramatically different from factory farming, particularly in terms of the welfare of the pigs. The animals are not packed and subsist on a diet largely free from commercial feed. There is also no compulsion to fatten the pigs quickly. Indeed, both the big and micro varieties of the small-ear pig are reared by the villagers and the former can grow up to a maximum of 100 kg for boars and 110 kg for sows (China Genebank, 2005), although the slaughter weight of these pigs is about 50–60 kg on average

Figure 6.5 Banana tree trunks used as (organic) feed for small-ear pigs

(compared with the country average of 78 kg). The micro variety can grow up to 25 kg (for boars) and 35 kg (for sows). The slaughter weight for the micro variety is about 15–18 kg. The slaughter ages for the big and micro varieties are nine and seven months respectively. While the physiology of the small-ear pig (for example, low prolificacy) presents distinct disadvantages compared with other commercial breeds, it is priced at about 30 per cent more than the latter. The price for each variety of pig ranges from 300 to 700 yuan ($45–105) with the micro variety fetching more per kilogram. As its name suggests, the selling point of the small-ear pig is precisely its relative diminutive size.

In Jiama we met the most successful farmer who managed to sell a total of 27 pigs in 2007. He was keen to expand his operation but feels hampered by the lack of space to enlarge his current pig shed. He explains: 'Here in the mountain, I cannot hope to have a big farm, it is impossible. We need to be on the plains to do that.' One Agricultural Department official frequently reminded us that this scheme was conceived as a supplement to farmers' income and not to turn them into full-time pig farmers. With only one sales point selling two to four pigs per day (depending on supply), the demand for the small-ear pig far outstrips supply and the meat is often sold out even before 9 a.m. The Agricultural Department has successfully marketed the small-ear pig as a safer, more delicious meat that is not only a pure breed but purely bred. As one official explains:

The consumers take to small-ear pig easily. They know how commercial pigs are bred and how contaminated with hormones and chemicals they can be. Small-ear pigs are more natural and safer. It is our indigenous pig. We don't need to do much advertising at all. Everything is done through word of mouth. The potential for this meat is great. Right now, if we have enough supply, we could even sell it to Kunming or overseas. It has a lot of potential.

The case study of a relatively remote region in China illustrates the need to understand the social cultures of meat production and consumption. In this instance, the poverty alleviation scheme of encouraging villagers to raise a niche, indigenous species of pig is strongly pivoted upon the distinction of such pigs in the first place. It furthermore capitalises on the changing consumer culture that increasingly places a premium on organically produced meat, even as this customer base is miniscule at the moment. The apparent success of the scheme nonetheless belies significant tensions and contradictions.

For example, many villagers do not yet treat the scheme as a *bona fide* income supplementary activity, with some we spoke ending up eating their own pigs or giving them away to friends and relatives as gift exchanges. In other words, there is a tendency, most likely because of the small-scale nature of the initiative and the fact that the sales of pigs have only marginally increased income, for the villagers to treat the pigs as livestock for personal consumption, social bonding and even as a status symbol. This is aggravated by the fact that no formalised contract exists between the villagers and the Agricultural Department on the number of pigs each household must (re)sell to the department. In sum, what this suggests is a subconscious resistance to the (extreme) commodification of livestock that is possible only because of the specific nature of the organic pig socio-economic system in the highlands of south-west China.

More importantly, the very organic culture of production could be under threat too. Believing the immense potential of this meat and recognising the limited supply coming from the villagers, the Agriculture Department has begun to consider rearing the pigs themselves on their own underutilised farms. In recent years, the department made an open call for investors to help expand the small-ear pig industry with an injection of 15 million yuan ($2.3 million). It is an ambitious plan that aims to capture 5 per cent of the high-end pig meat market in Yunnan province where demand is estimated to be 6,000 pigs per day. Hence, the planned daily supply of the small-ear pig is 300 per day – a drastic increase from the two to four that they had previously delivered. Along with this increase, the plan calls for up to 100 specialist butcher's shops to be set up in the major cities of the province, such as Kunming, Dali, Yuxi and Chuxiong.

Regardless of whether the Agricultural Department is able to attract investors and expand, the fact that such a plan is even conceived suggests that the intensification tendency of the livestock industry does not exempt hitherto organic and niche breeds. Furthermore, it provokes scepticism about the intentionality and integrity of the authorities in using niche livestock farming as an income diversification scheme for marginalised minorities. Should the planned 'industrialisation' of the small-ear pig be successful, it is likely to jeopardise the existing informal arrangement between the Agricultural Department and the upland Jinuo people, as well as transform the very nature of these pigs (for example, their higher welfare status). In that sense, a multifaceted scheme that aims to alleviate poverty and offer an alternative to industrial meat might unravel in the face of the relentless push for profits. That itself is a testament to the tenacity of the intensification logic where production is always susceptible to the pressures of profit generation. As opposed to tinkering with production methods, a more fundamental response against industrial farming is to just stop consuming meat.

Vegetarianism and Social Action

A study extrapolates that there are 75 million vegetarians of choice in the world and 1,450 million vegetarians of necessity (Leahy, Lyons and Tol, 2010). The latter are defined as consumers who are too poor to afford meat but would likely consume meat once their income level rises. Indeed, one of the greatest sources of optimism for meat producers is precisely the belief that with greater global wealth there will be a significant increase in meat consumption, driven by vegetarians of necessity. For example, the FAO has projected meat consumption in East Asia to rise to 650 kilocalories per person per year by 2050 (the corresponding figures in 1970 and 2000 were 100 and 400 respectively). Not eating meat, then, is first and foremost a function of economic status. However, it is also a marker of culture (Kalof, et al., 1999; Shukin, 2009) and religion (Donner, 2008; Nath, 2010; Miller, 2011).

Vegetarianism advocates have to constantly navigate, and to some extent overcome, these markers as they advance their goal of reducing meat consumption among the population. Put another way, they have to persuade non-vegetarians that vegetarianism is not just for people of particular cultures or religions. Conversely, they have to also convince others that consuming meat should not be linked with superior economic status. Hence, destabilising the cultural-religious marker of vegetarianism is imperative if activists want their message to reach more people. Indeed, adopting a meat-free diet is not easy for most people in the absence of the motivational and moral dictates of culture and religion. In that sense, vegetarianism advocacy is universally and unusually challenging because it is essentially a form of social action whose ultimate goal is to realise a considerable lifestyle change in the mainstream population.

Despite the ever-growing demand for meat, there has been a perennial countertrend to meat consumption. More than a hundred years ago, in 1908, the International Vegetarian Union was formed to specifically advocate vegetarianism worldwide. The first known vegetarian society − the British and Foreign Society for the Promotion of Humanity and Abstinence from Animal Food (the precursor to Britain's Vegetarian Society) − was formed even earlier in 1843. The two main reasons for establishing vegetarian societies and promoting vegetarianism are health related as well as to promote a more humane relationship with animals. Beyond health and animal welfare reasons, historically vegetarianism and food animals in many societies are markers of, among other things, culture, ethnicity, religion and even class status, as noted earlier.

Ultimately, a person's dietary choice is both personal and grounded in particular social and cultural structures. Similarly, as will be discussed, the willingness to partake of extreme forms of commodified meat, that is almost divorced from its organic fleshy form, is dependent in part on the consumer's sociocultural grounding. In what follows, then, we first contextualise vegetarianism advocacy before discussing the challenges of vegetarianism advocacy and the future of synthetic meat and other forms of meat analogues.

The proliferation of vegetarian societies around the world attests to the fact that activism and education on meat reduction and avoidance are ever-present. However, the specific politics of vegetarianism activism is played out differently in various parts of the globe. Large-scale events held to promote vegetarianism for health reasons are generally collegial and non-confrontational. For example, Meatout, a product of grassroots social activism, is held on the first day of spring in the United States to educate communities, friends and families to reduce meat consumption. First organised in 1985, the nationwide event aims also to persuade people that a vegetarian diet is more wholesome.

As alluded to earlier, vegetarianism advocacy is unique when compared with other socio-environmental movements in that its members are often expected to make material changes to their lifestyles, to a meat-free diet. Thus vegetarians and vegans also negotiate a very personal politics at a day-to-day level as when they are in social gatherings where food is served. In their frequent interactions with families, friends and colleagues on their dietary choices and explaining to them their reasons for adopting a meat-free lifestyle, vegetarians and vegans are essentially politicising vegetarianism in subtle ways. Put another way, they are exemplifying the fact that the personal/body is political and that being vegetarian is arguably a form of embodied social action which impacts on the immediate social sphere of vegetarians. Indeed, research has shown that people generally change their diets through personal social interaction with already committed vegetarians.

Beyond the individual and bodily scales, there have also been city-level efforts to promote vegetarianism through the symbolic declaration of a meat-free day (or veggie day) each week. The earliest place to adopt this was the Belgian city of Ghent where each Thursday is designated as a 'veggie day'. Other cities that have adopted a meat-free day each week include Cape Town in South Africa, Bremen in Germany and São Paulo in Brazil. Overall, cities and organisations in more than 35 countries have pledged a meat-free day (Meat Free Monday, 2014). While such gestures are almost impossible to enforce legally, it is to the credit of vegetarianism activists to have successfully launched such high-profile sociopolitical campaigns, amidst scepticism from various quarters, including the restaurant and meat industries. Indeed, there are signs that such initiatives, even if they can be criticised as merely symbolic, are being increasingly replicated in education institutions and private companies.

Another increasingly important space from which to engage in vegetarianism activism is the meat commodity chain where livestock producers have marketed meat consumption as a viable and enviable dietary choice. Animal advocacy groups like People for the Ethical Treatment of Animals (PETA) have continually exposed the cruelty prevalent in the meat production process through undercover activism (see Chapter 5). For advocates, such sociopolitical exposés highlight the fact that animal cruelty is integral to the modern production of cheap meat. The underlying assumption – which we question later – is that such knowledge can compel ordinary consumers to think of their culpability in the meat commodity chain. In so disrupting the consumption–production links through thoughtful reflection, activists hope that consumers might be persuaded to stop eating meat.

The International Vegetarian Union lists hundreds of affiliate vegetarianism societies worldwide and organises an annual World Vegetarian Congress (International Vegetarian Union, 2013). The justifications for adopting a vegetarian diet are universal and have always been predicated on three broad and interrelated concerns: health, animal welfare and environmental well-being. Each concern, however, has limits in furthering the vegetarianism cause. For example, while there is a broad consensus that consuming too much meat is bad for one's health (Leitzmann, 2003; Walker, et al., 2005), it is much more debatable whether eliminating meat completely from one's diet (as opposed to reducing meat intake) is the best health option for people. Indeed, the US Department of Agriculture (2015) recommends that our diet should be made up of 50 per cent fruits and vegetables, 25 per cent lean meat, poultry or fish, and 25 per cent grains, with suitable amounts of fat-free or low-fat milk, yoghurt or cheese. Nonetheless, it is still not uncommon to find vegetarian activists promoting a meat-free diet as the healthiest one (Compassion in World Farming Trust, 2004), and in so doing going against conventional views of a balanced diet. To be sure, the notion of a 'balanced diet' is itself very much politicised and contested to begin with (see Nestlé, 2013).

In any case, many vegetarians take the argument from health further to raise a broader question that is steeped in antiquity: are humans 'natural' omnivores or herbivores? Put another way, what activists are arguing is not just that a vegetarian diet is healthier but that it is also more natural (see also Piazza, et al., 2015). As elaborated later, this is one important way in which nature is selectively mobilised to frame the vegetarian cause and, concomitantly, intersects in interesting ways the question of what is moral and ethical.

The concern for animal welfare runs in a continuum, each point presenting limits and challenges to the vegetarianism cause. The first animal welfare concern relates to the way in which meat is produced in modern livestock systems. The latter, variously known as factory farms and CAFOs (see

Chapter 5), are often criticised for severely compromising the physical and mental welfare of animals (Ilea, 2009). The overall drive towards hyper-intensification of meat production is said to have commodified animals into mere inanimate objects in an intricate network of contract farming (MacLachlan, 2005; Neo, 2010). Others have pointed how such systems of production have exploited human labour as much as animals (Watts, 1994). Predicating a switch to vegetarianism on account of the cruelty of modern animal farming means that vegetarian activists are logically and morally compelled to acquiesce to the con-sumption of animals that are raised in a more humane and natural manner than CAFOs. For example, as discussed earlier, there has been a persistent niche in organic livestock farming that purportedly subjects animals to the lowest level of suffering (Alrøe, Vaarst and Kristensen, 2001; Ilbery and Maye, 2005; Neo and Chen, 2009). Furthermore, for most meat eaters, while it may be clear to them that subjecting animals to extreme and unnecessary cruelty is unnatural and wrong, there remains still a persistent cognitive dissonance that prevents them from effecting any real change to their purchasing or dietary habits (Smart, 2004). Indeed, Brock Bastian, et al. (2012) have argued that consumers precisely deny mental capacities to food animals so as to facilitate effective meat-eating beha-viour. Meatpackers facilitate such denial by parcelling meat in sterile packages and selling them in the comfort of a supermarket; they also shield consumers from the ugly sites and sights of factory farming.

The second justification for animal welfare states plainly that it is morally objectionable to kill: consuming meat is fundamentally wrong and unnatural. Thus, how well and comfortable meat animals are raised is not relevant because they are still ultimately killed for human consumption (Pluhar, 2010). Such a view towards animals can be in the first instance deeply influenced by religious beliefs (Miller, 2011). Hence, it is not so much a moral concern for animals per se that drives such vegetarianism than a direct religious imperative. Religiosity, broadly defined, is long recognised as an important ideological conduit that unites the vegetarian movement and the animal rights movement (Maurer, 2002, p.62). In that sense, vegetarians who believe that it is morally wrong to rear animals for human consumption are generally the most con-sistent and committed ones, with the least likelihood of reverting to consuming meat. Be that as it may, studies have shown that the vast majority of people who initially attempted a vegetarian diet did so for health reasons (Morris and Kirwan, 2006; Fox and Ward, 2007).

Finally, vegetarianism activists have also used the rhetoric of envir-onmentalism for advocacy. This is a fairly recent strategic framing although the environmental impacts of meat production are not unknown. As recently as 2002, Donna Maurer (2002, p.62) observes that 'the vegetarian movement overlaps less with the environmental movement than with the animal rights and health food movements'. To be sure, the landmark publication of *Livestock's long shadow* by the FAO catalysed international attention to the links between environmental degradation and meat production (see also Jarosz, 2009). The rationale for aligning the vegetarian cause with the environmentalist cause is thus understandable as the latter is an established moral-ethical concern recognised by most people. However, compared with the health and animal welfare justifications, appealing to the environment to persuade people to turn vegetarian is not straightforward in terms of framing and eventual results. Not least, it is plagued with the same kinds of problems confronting the success of the environmental movement and, as is discussed below, some activists also have reservations about the seeming 'devaluation' of animals in such a strategic framing that elides the materiality of animals.

While for heuristic reasons we have briefly discussed the three justifications for vegetarianism separately, in reality all vegetarian activism touches upon all three in various permutations and to different extents. Indeed, vegetarians are often motivated by different justifications at different phases of their lives (Fox and Ward, 2007). Regardless of the exact framings of vegetarianism and the individual beliefs of activists, as a collective movement the most important goal for activists is to persuade consumers that meat consumption is both unnatural and immoral. In the next section, we focus on the vegetarianism advocacy efforts in Taiwan and Singapore (two East Asian countries that are projected to have a significant increase in the consumption of meat) to show how vegetarianism is framed as a form of ethical food consumption. The data are collected from in-depth interviews conducted with advocates in both countries in 2012 and the ensuing discussion drawn in part from Neo (2015, 2016).

Framing Vegetarianism in East Asia

In December 2009 a representative of the Taiwanese parliament – the Legislative Yuan – filed a motion to debate the possibility of introducing a meat-free day for all government institutions in the country. Huang Chih-Hsiung, a legislator from the then ruling party, Kuomintang, argues that based on studies done by the World Bank and United Nations, the livestock industry releases up to '60 per cent of global methane emissions'; he further elaborates that not only will consuming less meat lead to better health, it will also decrease Taiwan's carbon footprint. His motion found bipartisan support with 12 other legislators. With the support of the president of the Legislative Yuan, a non-binding motion was overwhelmingly passed to encourage all government bodies that provide meals to their employees (including schools) to mandate a meat-free day every week. Although non-binding, the motion found the greatest support in the Ministry of Education and was an unambiguous success. As of early 2011 close to 90 per cent of the 3,517 schools in Taiwan mandated at least one meat-free day in the school cafeteria.

Environment

The motion raised by Huang was the outcome of the lobbying efforts of an informal group of civil society activists formed in mid-2009. Calling their grouping the Meatless Monday Platform (周一无肉日平台), core leaders include novelists, journalists, academics and environmentalists. The public campaign orchestrated by the platform was distinctive in the way it emphasised the environmental merits of consuming less meat and, to a lesser extent, the health benefits. It is especially remarkable for its relative silence on animal welfare and animal rights. This can be plainly seen in one of their most popular slogans: 'Go Veg, Love the Earth' (吃素, 爱地球). Indeed, their publicity materials and online magazines repeatedly highlight the environmental impacts of meat production and consumption against the refrain of loving and caring for the Earth.

When asked about this, Helen, one of the key advocates of the platform explains: 'The Taiwanese public understands when you talk about environmental protection; so if we say not eating meat is also protecting the environment, the message will be clearer. And I think more effective'. Mark, another advocate, responds to the same question by arguing that given the already high level of environmental consciousness in Taiwan, it is opportune to emphasise the environmentalist aspects of a vegetarian diet:

> Most people here are very aware of the need to protect the earth, reduce our carbon footprints, and we have very good recycling rates. In a way, you can say that we are being somewhat expedient, but the message is not as important as the final outcome ... that we need to let people see it is wrong to consume meat.

Evidently, the obvious limit of a meat-free day is that while it will reduce the amount of meat the typical consumer eats, the number of vegetarians might not increase at all. In other words, its impact in reforming the modern meat industry – through such public pressure and social change – is limited at best and inconsequential at worst. Even the advocates of the platform (all of whom are vegan) are cognizant of this fact. As Helen elaborates:

> Yes, of course our biggest wish is that there will be more vegetarians as a result of this initiative [Meatless Mondays], but this is something beyond our control and honestly speaking there is no way for us to know. But at least this is a start and it has to be seen as a small step towards something bigger in future. If it turns out that there is no sustained or discernible change in our dietary habits, then at least we can say that we have elevated vegetarianism in the public consciousness.

In the case of the platform, one can easily see how 'nature' and the environment are valorised to make persuasive the vegetarianism cause. Meat is being framed as an environmentally detrimental product: an immoral, unnatural object. This framing is strategic insofar as recent research has suggested that people are most willing to reduce meat consumption for moral and environmentalist reasons (Tobler, Visschers and Siegrist, 2011). We argue that while the

Taiwanese activists have tried to frame meat as an immoral object through its proven negative impacts on the environment and a more amorphous appeal to 'love' the Earth, it is not a strong enough moral reason to encourage consumers to give up meat completely. Hence, only valorising the environmental impacts of meat consumption is unlikely to compel consumers to undergo a sustained radical shift in their daily lives. A critical explanation is that, understood plainly, meat as a category of food product has no meaningful substitute (ethical or otherwise, but see the discussion below on meat analogues). Ironically, few would agree that the sole purpose of eating is about functional survival. To elaborate, if meat consumption is the functional partaking of the requisite nutrients, then meat and some non-meat foods can be substitutable. This in turn can help ease the shift from consuming meat to a vegetarian diet.

For the Vegetarian Society Singapore (VSS), activists feel that few positive results will be achieved by highlighting the environmental impacts of meat consumption, illustrating how framings are influenced by the difference in sociopolitical contexts. As the president of the society elaborates:

> I think Singaporeans in general are not that environmentally aware. Sure, you might argue that the younger generations have better environmental consciousness but I don't see how anything concrete and material will come out of saying things like 'Save the world, don't eat meat!' It is already an uphill task asking Singaporeans to recycle as it is! So I think we need a different approach.

To be sure, the VSS is a formal non-governmental organisation and hence quite different from the Taiwanese Meatless Monday Platform. The biggest distinction between the two groups is that the latter exists for the sole reason of ensuring the success of the Meatless Monday initiative while the former juggles many different vegetarianism-related projects simultaneously (for example, compiling a directory of vegetarian eateries in Singapore). However, they are both united in their goal of ultimately seeing more vegetarians in their respective countries, even as their framings differ due to their sociopolitical contexts.

Health

Dave, an active member of the VSS with years of advocacy, explains why the society has relied more on the health justification to persuade people to become vegetarians:

> We need to be smart about what we do and how we do it.... So we say that if you want what is best for you, in terms of your health and mental well-being, you should consider becoming vegetarian. In fact, we have been trying to work out something with the Ministry of Health.

It may not seem explicit, but what Dave is trying to do is articulate a notion of ethics that believes that in order to be ethical one needs to start with taking responsibility for one's body. This idea is not uncommon and can be seen in discourses on obese people and smokers: how they 'owe' it to themselves to lose weight (see Guthman, 2011) or quit smoking (Chapple, Ziebland and McPherson, 2004) so that they can become 'better' people. In this way, attention is shifted towards what it means to be a moral subject. However, to such rationalisations there also exist limitations, as Gregory, another VSS member explains: 'This is a dilemma because, as a nutritionist, I cannot say that a diet without meat is necessarily better than a diet with very little meat. I think if you want to sustain a long-term vegetarian diet, it has to be more than for health reasons'. His view is echoed by Helen when asked of the limitations of her platform's Meatless Monday initiative in the school system:

> It is not easy to convince Taiwanese parents that children can grow up strong and normal without meat. Our Meat Free Monday [in the school system] is good because we are taking things step-by-step. I believe that a vegan diet is the best for health and most natural but I also know that not many people will agree with me on this.

Similar to the environment narrative, the health justification is not able to convincingly demonstrate the superiority of a meatless diet. Not least, as mentioned earlier, health and nutrition institutions invariably recommend a diverse diet. While it is clear that most consumers should cut down on the amount of meat they eat, it is not the case that a completely meatless diet is distinctively superior to a diet which contains some, but not excessive, amounts of meat. In other words, vegetarianism activists are unable to frame the consumption of low quantities of meat as unethical and immoral, from a personal health perspective. Furthermore, even if such a framing were possible (and sidestepping the issue of the critique that a 'balanced diet' is a political, discursive construct), a libertarian objection would be that a person's body is her own to control and there is nothing immoral or unethical about subjecting oneself to health risks through smoking, drinking or consuming meat.

From a health perspective, vegetarianism cannot be sustained as a form of ethical food consumption. This is because, as mentioned earlier, there is no meaningful substitute for the excluding of meat in one's diet – after all animal protein is completely different in taste and texture to plant protein. Furthermore, the necessity of substituting cannot be established convincingly, against the refrain of a balanced diet. In that sense, the ethical framing fails not because consumers cannot see that it is better (in the ethical sense) to consume less meat; rather, the 'ethical depth' of consuming only vegetables is simply too shallow to effect the drastic lifestyle change of being a meat eater to a vegetarian.

The point about the 'naturalness' of a meat-free diet, briefly mentioned above, evokes both the notions of nature/environment as well as health/body.

In this narrative, becoming vegetarian is not only good for the environment and for our health, it is also argued to be the 'natural' thing to do. This is a recurring justification in both Taiwan and Singapore. As Mark explains:

> If you ask me, humans are not meant to consume meat. If we go back millions of years ago, we are gatherers. We are so afraid of wild animals, so how can we eat them? If we want to go back to Mother Nature, to our most original state (返朴归真), it is where we don't consume meat, we respect animals! The Earth is sick now, and if we do that it will heal naturally!

Mary, a member of the VSS offers similar views but relates them more to the body:

> I don't think the kind of meat we eat now can be considered natural. It is completely unnatural for us to eat meat now with all the toxins and hormones and whatnots floating in the production system. Our bodies are just polluted by all these poisons! Maybe if we go back to using our bare hands to kill animals, then it can still be considered natural. Otherwise, now the meat we are eating is making people more aggressive, more impatient because of all the unnatural chemicals in them. It is true. Believe you me.

Here, something that is unnatural is at the same time deemed as socially immoral and unethical. This is not an uncommon position given geography's longstanding effort to disrupt the nature–social dichotomy (Castree and Braun, 2001). Suffice to say, outside of the vegetarian social sphere, not many will agree that consuming meat is unnatural – as evinced by ever-increasing meat production and the normalisation of meat as part of a balanced diet. It is telling that quite a number of vegetarians have similarly expressed the unnaturalness of meat consumption even if official discourse used in their public campaigns often sidesteps this belief. If anything, this is a clear example of the multiple meanings embedded in ethical food framings, such as vegetarianism.

Animals

The plight of (food) animals figures least in the public advocacy efforts of Taiwanese activists and, to a lesser extent, in those of Singapore's too. Yet it is an issue that provokes the most discussion and insights in our interviews with these activists. This is clearly unsurprising given the litany of animal welfare abuses in the modern meat industry which these advocates are cognizant of (see Chapter 5). Table 6.1 below details excerpts of interviews from six different activists when asked the questions: How do you feel about animals in the livestock industry? How can we use animal welfare and animal rights to validate the vegetarianism cause?

Table 6.1 Excerpts from interviews, focusing on animal welfare

Name	Interview excerpts
Jacob (Singapore)	Sometimes we need to think for a moment. What kinds of lives do animals live? Do we want that kind of life? The global meat industry is most perverse. One good strategy is to get all those undercover videos [of factory farming] out to more and more people.
Mary (Singapore)	There is no way that what we are doing is sustainable. We cannot continue to demand more meat, cheaper meat and not expect animals to suffer more and more. The difficult part is to let people see that all this is unsustainable.
Pete (Singapore)	I feel strongly about taking the life of another living thing. Animals are sentient beings. It is as wrong to kill a human being or a dog or a cat as it is a pig, a chicken or whatever. Killing something so that you live is not right. And we can live without killing…. I often ask myself, why do so few others share my view? I think it is hard to push the animals' right [to live] message to people. Singaporeans won't buy it unless they've already bought it … you know what I mean?
Helen (Taiwan)	Livestock animals are such poor things. I know many people do not care about them. But that's because they do not think about these poor animals … I believe it is a question of whether humans can be humble and understand that being human does not mean we can do whatever we want to do. Kill anything that we want to eat.
Mark (Taiwan)	I think this talk of animal welfare misses the point. The question we should ask ourselves is do we need to eat animals to survive? I think prior research and my personal experience has shown that we do not. So why are we raising animals to be killed for food then? When did it start that humans think it is desirable to do so?
Loong (Taiwan)	For me, it is all about the welfare of the animals. Your motivation to become vegetarian must be strong and consistent. I have seen many people slip in and out of vegetarianism and this is because they do not really know why they are vegetarian, or the reasons they become vegetarian in the first place are weak. From my experience, the most committed vegetarians are those commanded by their religion to not eat meat and people like me who care deeply about the animals.

While the excerpts seem rather similar, two different positions are actually apparent. The first is a critique of the deplorable welfare standards in the modern meat industry (seen in Jacob's and Mary's comments), while the second argues plainly that it is wrong to kill animals for food (seen in the remaining four comments). As mentioned earlier, the first critique means that it is not wrong to consume organically raised animals that likely had the highest welfare accorded to them. However, because organic meat typically costs much more than conventional meat (up to five times more expensive), it is arguably not a meaningful substitute for it yet. In other words, organic meat is not completely substitutable for conventionally produced meat due to its inherently exclusionary higher prices. The implication of the second critique is that organic meat is not ethical because killing for animal meat – regardless of the process used to produce it – is morally wrong. To substitute conventional meat with organic meat does not negate the fact that a morally wrong death has occurred. The latter point is a constant refrain by many activists. For example, Mark explains at length:

> It is fine to have organic vegetables or fruits because the point is to save the environment, reduce pollution and all that. But I think it is laughable to talk about organic meat. Sure, organically raised animals might have better lives compared to their 'normal' counterparts in many aspects; but they share the most critical similarity. Death! How can eating organic meat be more ethical? I think it is a scam.

As noted earlier, notwithstanding their strong personal views about the critical importance of animals in influencing their decision to become vegetarians, all the activists acknowledged that to sell the vegetarianism message through the animal rights approach might not be fruitful. Instead, they have continually emphasised the health and environmental benefits of not consuming meat.

The preceding sections illustrate the challenges faced by activists to frame vegetarianism as a form of alternative, more ethical food consumption. By extension, it also explains the difficulties in enlarging the number of committed vegetarians globally. The use of nature and morality metaphors by the activists on both themselves as well as meat (consumption) shows how the 'ethics of ethical foodscapes' is 'ambiguous, slippery and consists of a number of interwoven layers' (Goodman, Maye and Holloway, 2010). For the Taiwanese, grounding their campaign on the premise of an ailing Earth demonstrates how Nick Clarke (2008, p.1873) characterises activist organisations; that they

> seek to persuade them [consumers] to take action, not by providing information, but by providing narrative storylines that acknowledge the complexities of modern subjectivity, while connecting them up to themes of inequality and exploitation.

The example of vegetarianism advocacy and the ways in which meat are selectively constructed by advocates illuminate the multiple forms of meat, beyond its function as a source of food. They also concomitantly serve as a refrain against the commodified production of meat that emphasises quantity to the neglect of the quality of meat produced and especially to the oversight of the lives of food animals. Nonetheless, we argue that the failure of activists to frame vegetarianism, through various discourses and justifications, as a form of ethical food consumption can be explained through the absent of a viable substitute for meat. To elaborate, in most of their framings (for example, environment, health and animal welfare), they are unable to convincingly demonstrate to the average consumer that there exists a meaningful substitute for meat. This is stepping aside the fact that some of the framings do not even suggest the need to omit meat from one's diet – as opposed to merely cutting down on meat. On the other hand, the only semblance of a meaningful substitute, organically produced meat, is one that actually sustains meat consumption and runs completely counter to their cause of increasing the number of vegetarians in the world. Confronted with the prospect of organically produced meat that exerts least harm to animals and environment alike, the activists' moral position has to be elevated to one which holds that the killing of sentient being is wrong. It is a moral position that demands the most of the would-be moral consumer.

Focusing on the substitutability of the food product that is allegedly unethical provides critical insights as to what ethical food and unethical food mean in relation to each other. It also explains the ambivalence of consumers towards certain forms of ethical food consumption. The case study of vegetarianism demonstrates the variability of ethical food consumption practices and why the informational gap narrative (that is, consumers are unaware of the cruelty of conventionally produced meat) and the cognitive dissonance model approaches (that is, consumers do not care enough) cannot sufficiently explain, in the first instance, what makes certain food more ethical. More educational information or deeper emotional connection are unable to effect concrete changes on the part of the consumers if the existence of substitutes cannot be realised. This ultimately is more of a practical, pragmatic impediment for consumers.

This discussion of the challenges faced by vegetarianism advocates demonstrates why a meatless culture/world is one which is hard to imagine and almost impossible to materialise (Emel and Hawkins, 2010). In sum, vegetarianism advocates struggle to frame meat as a form of unethical food in such a way that will convince a greater number of consumers to *give up* meat, as opposed to merely reducing meat intake. This is made worse by the political-economic power of the global meatpacking industry in, among other things, revolutionising the production systems of meat, and ultimately popularising and normalising meat consumption through lowering the prices of meat (see Chapter 3).

This latter point on the cost of meat is pertinent for several different reasons. As noted, although organically raised meat addresses in part the welfare concerns of meat animals, it fundamentally cannot achieve the goal of omitting meat from one's diet. Moreover, as a form of ethical substitute to conventionally produced meat, organic meat remains out of reach to all but the most affluent consumers due to its relative high cost. In other words, that organic meat, as a substitute for conventional meat, is at present an ethical product which can only be purchased by the well-off is problematic because it presents a counter-intuitive case where consumers claim moral superiority due largely to their ability to partake of an expensive ethical substitute.

Besides organic meat as an expensive and problematic substitute to the abolitionist cause of vegetarianism advocates, a more radical solution is 'artificial meat' (variously known as 'synthetic meat' and 'laboratory meat'). In a part publicity stunt, the radical animal rights group PETA offered a reward of $1 million for anyone who could 'make the first in vitro chicken meat and sell it to the public by June 30, 2012' (Digregorio, 2008). As highlighted in Chapter 1, synthetic meat will not be commercially viable for some years to come, even as further research will likely bring down costs and improve its similarity to the taste and texture of meat. In the next section, we explore other socio-moral challenges of synthetic meat, set in a future where their cost and taste become more palatable.

Cultured Meat and Meat Analogues: Resistance or Commodification Redux?

It must be stated at the outset that cultured meat and, to a lesser extent, meat analogues are underpinned by technology. A precedent of cultured meat is cloning, which has been undertaken commercially to reproduce breeding stock. In 2008 the FDA determined that meat from cloned cows, pigs and sheep was fit for consumption, although the USDA has called for a voluntary ban on cloned animal products entering the food networks. Other countries like Denmark and New Zealand have placed a national ban on commercial cloning of food animals.

To be sure, cloning is actually quite expensive and only marginally successful. A report published in 2010 by the EU states that cloning success was only 10 per cent in bovine animals and 5 to 17 per cent in pigs (European Commission, 2010). Common problems include the failure for placenta to develop and large offspring syndrome. From July 1998 to September 2009 some 575 cloned cattle were born in Japan of which 55 per cent died shortly after birth. Once born, the mortality of clones is much higher than regular animals because of cardiovascular problems, kidney and liver failure, immunodeficiency, irregular skeletal formation and other factors. Nevertheless, in the United States, where cloning is perhaps most developed, there are three companies that sell cloned animals to breeders – apparently 'hundreds of pigs' and the 'thousands of cattle'. ViaGen is one of the three companies and perhaps their most famous clones are of Delta Darlene, a longhorn from the Marquess Cattle Company in Texas. Many other countries host research and development institutes and private companies that clone as well. Avantea in Italy cloned the first horse as well as the first offspring of an adult bull.

Margaret Atwood, the Canadian writer, gets at the future of 'meat' products in her speculative bioengineering novel, *Oryx and Crake* (2003), by presenting the reader with 'ChickieNobs'. These are made from 'chickens' that no longer have heads but are hooked up to machines that provide the electrical and other impulses necessary for the bodies to grow but without the messiness of sentience or thought.

> What they were looking at was a large bulblike object that seemed to be covered with stippled whitish-yellow skin. Out of it came twenty thick fleshy tubes, and at the end of each tube another bulb was growing.
>
> 'What the hell is it?' said Jimmy.
>
> 'Those are chickens,' said Crake. 'Chicken parts. Just the breasts, on this one. They've got ones that specialise in drumsticks too, twelve to a growth unit.'
>
> 'But there aren't any heads …'
>
> 'That's the head in the middle,' said the woman. 'There's a mouth opening at the top, they dump nutrients in there. No eyes or beak or anything, they don't need those.'

Cultured or in vitro meat is a step beyond ChickieNobs. There is no actual body that metabolises feed but cells that are fed instead. While cultured meat is even further off commercially, it is not exactly in the realm of science fiction. As alluded to in the opening paragraph of the book, synthetic meat may have become the vanguard of advanced biotechnical food animal science. In another venture to produce alternative meat, scientists working for the company Modern Meadow are experimenting with 3D printing of meat tissues. These researchers, funded by a private financier (in this case Peter Thiel, an investor in PayPal, Facebook and hedge funds), hope to develop 3D cellular sheets composed of pig cells. The intention is to mature those sheets into muscle tissue with electric stimulation inside a bioreactor (which should help give it the desired texture). It is roughly the same recipe for producing 3D human tissue and will benefit from advances in human tissue bioengineering.

At its simplest, synthetic meat presents a radical way to sidestep the moral dilemmas of killing food animals for meat while continuing to supply the growing, insatiable taste for meat. In abolishing meat yet providing opportunities to consume 'meat', synthetic meat presents an opportunity to have one's cake and eat it. Research in synthetic meat technology has intensified in recent years, and it is timely to problematise the ethical, social and political-economic dimensions of this development.

Patrick D. Hopkins and Austin Dacey (2008, p.582) detail the various emergent scientific advances that indicate a possible future of synthetic meat where 'a laboratory could produce meat through a technological process; real steaks, real prime rib, real chicken breasts, real veal, grown in a lab'. Such scientific breakthroughs include 'scaffolding' where 'skeletal muscle cells can be grown on small beads or mesh suspended in growth medium'. Other notable technologies include organ or 3D printing, as well as nanotechnology. As Hopkins and Dacey surmise:

> Technologies ranging from the actual to the speculative promise a variety of ways to create real meat without killing animals.... Though still commercially infeasible at the moment or in some cases techno-logically infeasible for several years to come, the point is not to be dis-tracted by the fact that we cannot yet make use of these technologies but rather decide whether we should support the development of these technologies.

In other words, given the pace of technological change in the livestock industry, it is worthwhile to consider the potential and limits of synthetic meat at this juncture. At its most optimistic, the technologies of synthetic meat production ensure that no animal would be killed in the process. Mark Hawthorne (2005), drawing on the famous animal rights ethicist Tom Regan, argues that if 'no injustice is done in the procurement of the cells (used to produced synthetic meat), then it's more difficult to lodge an objection rooted in respect for animal rights'. However, we take a more nuanced stance with regard to this point, by valorising the notion of 'natural'. For animal rights activists, the questions of 'harm' and 'natural' are quite different things. In other words, that which is harmless does not necessarily mean that it is natural. Hence, even if extracting cells from live animals to make meat does not harm the animal, one might still question whether it is natural. Moreover, many animal rights activists are wont to argue that similar to that which is harmful, that which is unnatural should be made illegal too. A good example would be cetaceans held captive to perform at marine parks. Even if the animals are not physically harmed (which itself is a highly questionable assumption), anti-captivity activists often base their objection on the unnaturalness of the captive lives of cetaceans (Neo and Ngiam, 2014). Similarly, we agree to some extent that artifice or unnaturalness is morally problematic in and of itself. To conclude otherwise is to tacitly accept the idea that all that matters is visible, physical harm and that the intrinsic well-being, integrity and especially the authenticity of the animal are secondary. That said, ideas of nature and what constitutes as natural are clearly evolving and not ontologically fixed (Castree, 2005). Or, as Hopkins and Dacey (2008, p.587) put in more blunt terms, an objection based on the naturalness of synthetic meat can be reduced to the 'yuck factor'.

For them, drawing on the feeling of disgust to oppose synthetic meat is problematic for two reasons. First, food is notoriously rooted in cultural practices, making the definition of disgust spatially arbitrary at times (as the example of horsemeat consumption described in the second chapter alludes to). Second and more important, history has shown that people might not continue to feel disgust 'when educated about or familiarised with a new process (Hopkins and Dacey, 2008, p.587). Indeed, the kinds of confined animal feeding operations that have proliferated and become increasingly known to (and accepted by) many consumers would have been unthinkable mere decades ago.

Hopkins and Dacey's defence of synthetic meat can be recast differently when we consider critically the form and function of meat. The technoscience of synthetic meat is principally concerned with replicating the function of meat in terms of its taste and texture. Hence, deeper debates about the form of meat and its attendant meanings that fall outside of its functionality are not deemed essential. Yet, as we have argued earlier through the example of religious slaughter, both the form and function of meat are critical to the average consumer, albeit to different degrees.

To be sure, the functionality of meat presupposes its safety. In that sense, a minimum quality (pertaining to its safety) of synthetic meat is imperative. Supporters of synthetic meat are likely to sidestep this issue by placing their faith and trust in the scientific regulatory system to maintain safety. But as we have noted earlier, issues regarding food safety in the modern meat production system cannot be seen in isolation from the broader political-economic reality represented by, *inter alia*, large meatpacking companies, particularly the latter's ability to influence research and regulators (see Chapters 2 and 3). At the opposing end, anti-meat activists have long campaigned on the premise of the long-term, often hidden, negative health impacts of meat consumption. Such contestations are likely to play out in the case of synthetic meat should it eventually become widely available. Recalling our use of governmentality, the case of synthetic meat essentially allows us to see how formal rules and social-cultural norms (which arise from specific institutions) construct objects and how these are subsequently internalised by consuming subjects. The governance hence extends to both the object that is consumed as well as the consumer. Ultimately, governmentality also acts to purposefully produce knowledges and technologies which alter the 'choices, desires, aspirations, needs, wants and lifestyle of individuals and groups' (Dean, 2010, p.20). This can no more clearly be seen than in the likely future development of synthetic meat. The latter essentially problematises the dichotomy of production and consumption as well as form and function of a particular type of meat.

Returning to the link between meat production/consumption with carbon emissions, one would have to agree with Samuel Randalls's (2011, p.127) assessment that the debate on climate change parallels that of synthetic meat: it 'is as much of a cultural, moral issue as a technical challenge'. We might add that it would eventually become an advocacy issue as well, for when the day arrives that synthetic meat can be practically produced for the masses it would have to be accompanied by extensive campaigns to draw consumers to it. In the final analysis, the synthetic meat of the future will profoundly and fundamentally transform our understandings of meat, technonature and human–animal relations.

Meat Analogues

Despite their revulsion at the taste of meat, for vegetarianism advocates the recreation of the taste and texture of meat in other food sources is a welcome development and there are presently cheaper ways to achieve this than producing synthetic meat. This is a biting irony to some extent, and in Britain alone the value for the analogue meat market is estimated to be nearly £800 million ($1,148 million) in 2012 (Taylor, 2012). Nonetheless, there is no consensus, even among vegetarians, that meat analogues are *bona fide* substitutes for meat. As a non-activist Taiwanese in his fifties, who had been a vegetarian for about two years at the time of interview, explains:

> I feel that we are already very good when it comes to vegetarianism cuisine. We have all kinds of mock meats [meat analogues] that we use. But honestly speaking, they are not the same, I know what meat tastes like and these mock meats usually taste quite different.

Suffice to say, the scientific development of meat analogue is ever advancing. The Dutch company, The Vegetarian Butcher, is a self-referential, self-aware progressive company that sells meat analogues. And it is all the more newsworthy for being based in a largely meat-eating country. They describe themselves as such:

> After a ten-year search, Jaap (founder-owner) developed and found innovative meat substitutes with a spectacular bite and texture. With top chefs, he added the flavor and experience of true meat to the products. In addition, he saw a lot of potential in the protein rich and organic lupine from Dutch soil. Together with concept designer Niko Koffeman, chef Paul Bom and a devoted team, he is working on a big transition from animal to vegetable meat. Their ideal is to have meat enthusiasts experience that they don't have to miss out on anything if they leave meat out of their diet for one or more days. The ambition is to become the biggest butcher in the world, in a short time.
>
> (Vegetarian Butcher, n.d.)

The online shop features tantalising pictures of products such as tuna, beef strips, smoked bacon, chicken chunks, meatballs and shawarma. The meat analogues are named exactly as they are supposed to taste, feel and look like. Such meats obviously do not involve the actual killing of food animals, as with the argument for the production of synthetic meat. The key difference is that the taste and texture of in vitro meat are theoretically possible to match to the exact taste of meat. Elsewhere, meat analogues have gained even more traction. The Los Angeles-based company Beyond Meat has since 2009 actively researched on replicating the taste and texture of chicken and beef. Its two headlining products – Beyond Chicken and Beyond Beef – have garnered much more favourable tasting reviews (compared with synthetic meat) by food critics like Mark Bittman who writes: 'You won't know the difference between that [Beyond Meat] and chicken. I didn't, at least, and this is the kind of thing I do for a living.' Similarly, the chef Alton Brown pronounced Beyond Meat products as 'more like meat than anything I've ever seen that wasn't meat.' Since 2013 the American national grocery chain Whole Foods has started to carry the Beyond Chicken range of products.

Conclusion

In earlier chapters, we highlighted the costs of meat consumption to the consumer, food animals, the environment and the workers drawn into the global and/or modern meat production system. This chapter details two responses towards a growing meat market that we have taken pains to show is deeply problematic. Put simply, the first – organic farming – is driven by the production end of the equation and remains ever susceptible to the intensification logics of the global meat complex. Moreover, in catering to a select group of consumers, it might not significantly alter the hegemonic system of meat production at the global scale. The second – vegetarianism activism – is driven by the consumption side of the equation. Eliminating meat from one's diet or at the very least reducing the quantity consumed would destabilise and in the most extreme scenario dismantle the meat production system. However, as we have also shown, anti-meat activism in practice faces significant challenges. That the long and virtually unbroken history of vegetarianism exhortations has not resulted in extensive changes in global consumption habits attests to the tenacious ways in which a meat-consuming culture has spread across the world and taken hold of it. Joshua Frank (2007, p.319) argues that meat consumption is a 'bad habit' that is hard to give up. In addition, with the production side sustaining the meat eating habit, consumers are often locked in when it comes to their dietary choices. Modest breakthroughs in reducing meat consumption in some places are often more than offset by dramatic increases in other places. For example, while there has been a drop of 9 per cent in the consumption of meat in the United States from 2007 to 2012 (due to faddish low-meat diets and concerns over food safety), everywhere else the middle class and urban dwellers are eating more meat, particularly in developing countries (Heinrich Böll Foundation, 2014, p.8). Short of an improbable jump in the prices of meat or a global catastrophic zoonotic disease outbreak, it is increasingly unlikely that advocacy can make significant dents in the consumption patterns of meat at a global scale.

More optimistically, abetted by science, the growth and progress of meat analogues are showing encouraging signs, signalling a way forward to substantively modifying consumer demand for meat.

7 Conclusions

Introduction

In the preceding chapters, we have tried to illustrate the complexity of the meat industry, highlighting salient features across its entire production and consumption chain. The chapters are organised broadly in three different parts. In Chapters 1 to 3 we explain conceptually and detail empirically the spread of a particular model of livestock farming that is predicated upon continuous intensification, focusing on the fundamental reasons that allow for its extension. In so doing, we also take care to avoid generalising this broad trend by highlighting countervailing developments against intensification. Yet our conclusion is unambiguous: the spread is certain, even if it moves unevenly across the world. This certainty is explained by the way in which the architecture of industrial meat is able to govern and persuade actors along the chain of the inevitability and desirability of intensified production. In other words, we show how intensified production has been normalised through a critique of the political economy of livestock industry (Chapter 2) as well as the role of science and technology in the normalisation process.

In Chapters 4 and 5 we discuss the impacts intensification has on food animals (Chapter 5), workers and the environment (Chapter 4), since the intensification of production is not self-evidently bad in and of itself. Not least, such impacts are often hidden from the view of the ordinary consumer. Concomitantly, we also evaluate the various initiatives undertaken by governments and the industry to (self)-regulate and ameliorate the environmental and labour fallout of intensification. We conclude that the resilience of the global architecture of industrial meat and the way it garners the complicit support of the regulatory bodies suggest that in the foreseeable future the status quo will persist. In other words, ever-increasing demand will see an expanding supply made possible by more precise ways of intensified extraction of animal bodies. This forbidding conclusion then leads us to Chapter 6, which looks at how non-governmental organisations, 'enlightened' companies and ordinary citizens can dent and perhaps change this apparently inevitable trajectory.

It should be clear that *geography* underpins the emergence, spread and resistance towards industrial livestock farming. The global hegemony of the industrial meat complex is never homogenous in the way it touches down in specific places. A well-known quotation from Claude Lévi-Strauss's (1963, p.89) *Totemism* states: 'We can understand, too, that natural species are chosen not because they are "good to eat" [*bonnes à manger*] but because they are "good to think" [*bonnes à penser*]'. Although he was trying to explain specifically why animals represent the totem, it nonetheless unwittingly, through clever word play, illuminates two key ideas running through the book. The first obvious interpretation and idea is that animals are not just useful/good for consumption, they are also useful to think about, in terms of the broader reality that humans and non-human animals cohabit. We have illustrated this broader reality in various ways, grounded in various places throughout this book, thereby valorising the importance of geography. For example, the cultural politics of pig production in Malaysia shows how the idea of a food animal can range far beyond its functional role as sustenance. Similarly, the example of organic meat production in rural China draws on idea(l)s of nature, naturalness and moral consumption that enliven food animals. In both these places and examples, we look to food animals less as being food for humans than what they can tell us about human nature and society as well as the roles they play in (re)shaping human identities and ways of life.

On the other hand, running contrary to this latter conceptualisation of food animals as beings with agency and meanings is the second, more subtle and less popular, reading of Lévi-Strauss's statement which interprets the word 'good' to mean product or commodity. It renders animals as objects, devoid of meaningful lives and existence beyond their functional role as meat for human consumption. The most significant repercussion of the modern factory system is to have normalised the commodification of food animals as products to the vast majority of consumers. Such normalisation makes it ever more challenging to interject the lives of food animals with meaning and empathy so as to build better futures for them.

Magda Stoczkiewicz, director of Friends of the Earth Europe, has emphatically stated that 'the current industrialized and corporate-led system is doomed to fail. We need a *radical overhaul* of food and farming if we want to feed a growing world population without destroying the planet' (cited in Heinrich Böll Foundation, 2014, p.7, emphasis added). The imperative question to ask now is by what means can such a radical overhaul be achieved?

Softly-Softly and the Resilient Governmentality of the Meat Complex

One of the means through which governmentality materialises is the economic logic of a Fordist regime. As highlighted, in such a regime, in the interest of efficiency, production tasks are divided into minute detail, enabling goods to be mass-produced at lower prices. For the livestock industry, mass production has led to environmental ramifications at scales that were unheard of as recently as 30 years ago (see Chapter 5). The production of meat has continued to be predicated upon increasing productivity and standardisation. For ease of transportation, slaughtering, packaging and consumers' perennial demand for health and convenience, livestock animals in 'modern farms' are reared to precise requirements (Ufkes, 1998). Often, the animals that are passed through the contract production system are owned by the big processors and contract farmers merely hired to process the animals at different stages (see Chapter 3). Increasing numbers of farmers are adapting to such 'modern' production and organisational methods (for example, a contract system of farming and highly mechanised and circumscribed modes of production) introduced by established and powerful meat companies. In many cases, farmers have accepted such changing ways of production because of the recognition that the desire for consumers to have 'standardised meat' can only be met by the expertise afforded by large meat processors and the systems and technologies they possess. Commodification and intensification are thus normalised as an inevitable development to ensure economic viability. This is a mental and practical change to *production* norms that is governmentality par excellence.

In other words, even if one agrees that a 'radical overhaul' of food and farming is imperative, given the resilience of the hegemonic mode of production it appears highly unlikely that a revolutionary overhaul is on the horizon. Rather, the system needs to be ambushed from all sides, albeit in a softly-softly manner.

As Emel and Hawkins (2010, p.35) argue, 'our food choices at the institutional level are already highly constricted, especially as they relate to the political and economic clout of the meat and dairy industries'. They highlight the usefulness of institutional-level change that is gradual and incremental (citing the Meatless Monday campaign as a useful example) and driven by motivated individuals. As Emel and Hawkins further explain:

> Critics on the left might argue that individual consumption is not a route to changing structural modes of production, that in fact it reproduces neoliberal subjectivity ... They argue that it is absurd to think that we can change the system 'one meal at a time'. But we think that that is exactly how the system gets changed.

Political actions like these should aim to destabilise the meat industry by targeting the common rationalisations consumers make when choosing to eat meat. These revolve around seeing the consumption of meat as *natural*, liking the taste of meat and believing that it is necessary to consume meat (see Chapter 6). Above all, most consumers view the consumption of meat and (by extension) the production of meat as *normal* (Piazza, et al., 2015). The denormalisation of this perception cannot be achieved without dismantling the architecture of the global meat complex by breaking its incessant governmentality.

Geographies of Meat: The Missing Pieces and Future Prospects

As we indicated at the start of the book, we do not attempt a comprehensive sweep of the empirical and conceptual significance of the geographies of meat – that would be too onerous a task and one that might not be entirely fruitful. Instead, we have chosen to highlight what we see as the more illustrative issues relating to the geographies of meat through a combination of detailed empirical case studies as well as more general discussions. In this section we flag some other issues of interest and think about the future directions of meat studies in geography specifically as well as social sciences and humanities in general.

We reiterate three other issues that have been mentioned to various degrees in earlier chapters but bear repeating. As noted earlier, work safety and monetary compensation leave much to be desired in the meatpacking industry in general. Donald D. Stull and Michael J. Broadway (2004, p.75), in their engaging expose of the North American meat industry, note that 'reported injury and illness rate for meatpacking was a staggering 26.7 per hundred full-time workers in the late 1990s' – a figure that was three times more than that for manufacturing industry. As for wages in the meat and poultry industries, they show that a significant number of workers require federal assistance (in the form of medical aid and a lunch programme for their children) despite working full time. This means that not only are meatpacking companies paying less than a living wage, it also suggests that in effect the industry is indirectly subsidised by the wider society (through state and federal assistance schemes as well as charity) to reduce the prices of meat and depress the wages of workers. While research on labour welfare issue in the livestock industries of developing countries is scarce, one can hardly imagine it to be better than the situation in more developed countries.

Elsewhere, we have detailed more comprehensively the myriad environmental impacts of the global meat industry (see Emel and Neo, 2011). We have also specifically highlighted the inordinate risks carried by CAFO workers to various diseases in Chapter 4. The risks are however not limited to the confined spaces of CAFOs. In recent years, zoonotic diseases – infectious diseases that are transmitted from animals (sometimes indirectly through a vector) to human beings – have become a significant public health concern. Some zoonotic diseases like rabies have a long history and have not had any significant, fatal outbreaks in years. Others like malaria continue to afflict many regions in the world. Nonetheless, zoonotic diseases like rabies and malaria are not direct outcomes of the production and consumption of meat. However, others like the Nipah virus, *streptococcus suis*, BSE and bovine tuberculosis have all had their roots traced to the modern meat industry (see Chapter 2).

Geographers have, of late, begun to study the sociospatial aspects of zoonotic diseases and how these specifically impact on the lives of humans and food animals, as well as changing the structures of meat production (Anderson and McLachlan, 2012; Atkins and Robinson, 2013). Steve Hinchcliffe and Kim J.

Ward (2014) argue that the goals of eradicating or minimising zoonotic diseases 'open up political space for exploring an alternative politics of life'. This is in contrast to a corporatist and unyielding regulatory approach to disease control that aims at homogenising the differences in farming practices and obliterating the situated knowledges farmers have of food animals under their care (see also Allen and Lavau, 2014). Gareth Enticott (2008) has also demonstrated how biosecuritising meat production engenders contrasting and conflicting knowledges of food animals held by the regulators and farmers, resulting in negative welfare consequences among food animals, farmers and other 'pests' like badgers. To be sure, the practices of regulating food animal health is evolving, and Enticott (2014) has most recently argued that, with regard to bovine tuberculosis, there has been a gradual erosion of government control and clear disparities in disease diagnosis between vets in the private and public sectors. He shows how regulatory structures sometimes depart from standardised testing and control in the face of the complex realities of animal production. As with the issue of labour welfare, the study of the sociospatial impacts of zoonotic diseases has been selective in its empirical focus, with much more analysis addressing these issues in the context of more developed countries than the rest of the world. This is a research lacuna that has to be addressed.

In Chapter 6 we speak of substitutes to meat, such as meat analogues, that may contribute to a reduction in demand for food animals. If we view meat as animal protein, then there exist other possible substitutes from which consumers can obtain animal protein – the most important of which is fish. While there has been a wealth of research on fish, we have made the decision to sidestep this in the book in the interest of space. Both wild caught or farmed fish are consistently the most important source of dietary protein in many parts of the world. Indeed, the study of fisheries in the social sciences has burgeoned in recent years, with a growing interest in farmed fisheries (or aquaculture), which is analogous to livestock farming and faces similar issues and challenges. For example, there is a significant corpus of research which is tangentially political-economic. This focuses on issues of certification, health and value, on the one hand (Hall, 2010; Bush and Duijf, 2011; de Vos and Bush, 2011; Mansfield, 2011), and the industry's impact on the environment and rural development on the other (Schurman, 1996; Salmi, 2005). Straddling both these broad thematic foci are works that explicitly look at the role of institutions and market-oriented policies in shaping fisheries (Barton, 1997; Mansfield, 2006; Bush, 2010). Economic geography, on the other hand, sees aquaculture as a growing economic sector and is chiefly interested in unpacking the political economy of aquaculture companies (see Lim and Neo, 2014).

From an economic perspective, aquaculture provides a livelihood and income for nearly 16.6 million fish farmers globally, with 97 per cent of them concentrated in Asia (Food and Agriculture Organization, 2012). In poverty-stricken regions like Vietnam's Mekong Delta, indigenous fish farmers are able to raise and market their produce to wealthy consumers in developed countries with the help of the state, international institutions such as the Asian Development Bank and transnational capital (see Chapter 3 for a similar account in the livestock industry). Thus, fish are not only a valuable food protein for the farmers and their community but also a commodity that generates export earnings (Duval-Diop and Grimes, 2005) and hence a driver of rural change. These are all narratives that ominously echo the livestock industry.

However, there are distinct differences between fish meat and livestock. For one, the animal welfare concerns of farmed fish are less extensive than livestock animals. This is in part due to the conventional understandings of the physiology of fish, as opposed to food animals like pigs and poultry. Unlike the acknowledged sentience of food animal mammals, fish are less commonly believed to exhibit the range of emotions that would compel consumers to consider their welfare. Also, as capture fisheries are commonly thought of as common pool resources, a particularly popular research theme on capture fisheries analyses the effectiveness of community-based management approaches to fish stocks (Cheong, 2005; St Martin, 2006; Abbott, et al., 2007) – a line of enquiry that is seldom seen in either aquaculture or the livestock industry.

Nonetheless, in the final instance the aquaculture industry will likely mirror the livestock industry in becoming more widespread, intensive and internationalised. To this end, Christina Stringer, Glenn Simmons and Eugene Rees (2011) illustrate how the New Zealand post-harvest fisheries industry has undergone a transformation in which much of its harvested fish (not limited to those from aquaculture) are outsourced to China for further processing before such products are re-exported to the final consumers (in or outside of China). This arguably erodes the country's fish-processing knowledge base and its 'industrial commons – the collective R&D, engineering, and manufacturing capabilities that sustain innovation' (Pisano and Shih, 2009, p.116). Moreover, as with the livestock industry, the broader relationship between historical sociopolitical relations and their lingering impacts upon the development of the fishery industry and its related institutions is also under-researched.

Besides fish consumption, an emergent area of research on meat is the consumption of 'exotic' or taboo meats. This line of research echoes several themes that we have addressed in earlier chapters, like cultural politics, animal welfare and food safety. Hence, the consumption of cetaceans (Chang, 1978; Kalland, 1992; Fielding, 2013) and non-conventional meats like dogs (Wolch, Griffith and Lassiter, 2002; Wu, 2002), kangaroos (Hercock and Tonts, 2004), bushmeat (Fa, Peres and Meeuwig, 2002) and insects (Chen, et al., 1998) is a fertile research area that deserves more attention. The consumption of bushmeat in particular engenders other pertinent issues such as development and forest clearance which affects the supply of bushmeat to consumers (Fa, Currie and Meeuwig, 2003). The availability of bushmeat is a key determinant of food security in many rural places as well as for their household economy (Kümpel, et al., 2010), even as its consumption has been discussed concomitantly with the rise of novel forms of zoonotic diseases (Alexander, et al., 2012). Nonetheless, others have cautioned that the increased consumption of bushmeat might upset local ecosystems (Milner-Gulland, Bennett and the SCB, 2003).

Meat consumption and, more broadly, the intake of animal protein are thus deeply entrenched in different cultural norms across the world. Why some meats are more popular in one place compared to another has to do with the social-cultural roots of meat consumption. We argue that as a cultural practice – and here we define culture broadly to include concepts like gender and religion – meat consumption is for the most part spatially differentiated. Furthermore, when the cultural practices of meat production and consumption are linked to religious norms in particular, we are likely to encounter significant resistance to any changes to these practices, as well as rigorous debates over these practices. We address the issue of culture at various points in the book but more work could be done in this respect and it bears some reiterating here.

To be sure, such social cultures of meat are interesting detours from the economic commodification of meat, and we remain critical of a commodification process that strips food animals of their myriad meanings. The idea of purity is pertinent in food consumption in general and especially in meat consumption. Purity, in the anthropological, allegorical sense used and made famous by Mary Douglas's (1966, p.163) seminal work, *Purity and danger*, is the 'enemy of change, of ambiguity and compromise'. It is hence in part an ideology that pervades multiple facets of life. Yet conventional understanding of purity, when applied to the meat sector, can simply mean a desire for more 'wholesome' foods, as opposed to modern destructive means of food production. It is also used more metaphorically to refer to meats that are not taboo. Perhaps with the exception of chicken, most other meats are eschewed by a not insignificant number of meat eaters around the world. These meats include relatively mainstream examples such as beef (which is avoided by many Hindus and some Buddhists) and pork (which Muslims and Jews do not consume). In addition, there are some other meats that are consumed by a comparatively small number of people but are frowned upon by the larger population. We briefly look at the examples of dog meat, horsemeat and dolphin meat to illustrate the cultural underpinnings of particular meat consumption preferences and the opposition to such consumption.

A food taboo is predicated upon the idea of purity and the preservation of such purity involves boundary making and boundary enforcing, among other things. As Carolyn Rouse and Janet Hoskins (2004, pp.236–7) sum up succinctly:

> Food taboos exist simultaneously as a method for blending the physical and the moral; as a form of social control or a way of delineating order; as a way of reducing ambiguities; as a way of embodying resistance to disintegration (personal and social), and as a method for ascribing sacredness.

To put it another way, food taboos (or, more generally speaking, food avoidance) shape consumer identities and invocations of purity, and are often attempts to create and sustain a boundary between us and the other, most often seen in discourses of racial purity (Horwitz, 2001). Taken to the extreme, such 'othering' and boundary drawing become a discursive legitimisation for genocide – perhaps the most infamous example of which was Nazi Germany (Scales-Trent, 2001; Terzić and Bjeljac, 2014).

People who consume taboo meats like dog or dolphin typically defend themselves in two ways. The first is to argue that one is free to choose one's own dietary preferences. The second is that of a cultural mandate. In other words, one may argue that the consumption of such meats (seemingly taboo from the perspectives of many others) is an expression and extension of one's culture. Hence, those who object to such consumption practices are often accused of being cultural imperialists. *The Cove*, an Academy Award-winning documentary released in 2009, is a riveting tale that traces the killing of dolphins for meat in the coastal town of Taiji, Japan. Proponents of dolphin meat consumption in the area defended their dietary choice with an appeal to their cultural heritage. This is set against the argument of detractors who insist that the cruel slaughter of such sentient, intelligent beings is immoral and no civilised 'culture' would have allowed this practice.

The same cultural mandate argument is recast in the consumption of dog meat. In parts of Asia, including China, Vietnam and South Korea, some of the population historically consumed dog meat (Oh and Jackson, 2011; Avieli, 2011). In recent years, such consumption practices have been racialised as barbaric and cruel in other places (Elder, Wolch and Emel, 1998; Wolch, Griffith and Lassiter, 2002). The growing (countervailing) trend of keeping dogs as pets among the middles classes of these countries has produced localised objections to the consumption of dog meat, riding on the perennial protests by global animal welfare groups. The annual summer dog meat-eating festival in the town of Yulin, in Guangxi, China has seen increasing protests by dog lovers all over China (*Oriental Daily News*, 2015). Given this, one needs to be cautious about generalising the dog-eating habits of, say, the Chinese when it has erstwhile been a localised consumption practice. More importantly, it is likely that the cultural defence for such consumption will grow increasingly untenable and dog meat eaters will be increasingly marginalised in the near future. However, the politicisation of dog meat consumption in Yulin, and especially the way local people are increasingly demonised for it, can ironically sometimes valorise local identity to the detriment of the cause. When interviewed, a Yulin resident remarked: 'I actually do not eat dog meat, but because of the way they [the activists] scold us Yulin folks, this year I deliberately came to attend this festival' (*Oriental Daily News*, 2015).

In 2013 a scandal reverberated across Europe where processed food labelled as 100 per cent beef was found to have contained traces of horsemeat. In the worst case, the meat in a frozen beef lasagne product was actually made up of 100 per cent horsemeat and not a single gram of beef (European

Commission, 2014). Beyond the obvious consumer outrage over mislabelling, there were also deeper culturally rooted reasons at play. In Britain the consumption of horsemeat, or hippophagy, is deemed to be a revolting act by most consumers, even as neighbouring countries such as France and Belgium have had an unbroken culinary history of consuming horsemeat (BBC, 2013). Explaining why are there such regional divergences in horsemeat consumption is beyond the scope of this section. Suffice it to say, the divergence has occurred even though these countries have arguably similar cultural views on horses (for example, as beasts of burden, war horses, etc.). Such positive attributes generally preclude their consumption but, as can be seen here, such an outcome is not guaranteed. This suggests much more complex, culturally rooted reasons at play that continually challenge the universalist vision of justice for animals assumed by activists.

Nonetheless, in the final instance, there remains space to normalise other forms of taboo and exotic protein source like insects (Nel and Illgner, 2000). Some 80 per cent of a cricket can be consumed, compared with just 40 per cent of a cow and 55 per cent of a pig (Heinrich Böll Foundation, 2014, p.59). This is surely, in the jargon of food animal science, 'yield per animal' par excellence?

Conclusion

It should be clear by now that the geographies of meat are differentially intertwined (through time and space) with culture, political systems, economy and technoscience, across scales. From the meanings of meat through to the final act of consuming food animals, the experiences are highly varied from the standpoints of humans (as consumers or workers) and animals (as food to be consumed). Yet, what we see as an inevitable trend should be obvious enough: the cumulative commodification of food animals where their very essence and bodies are governed and disciplined, underpinned by (global) capital and consistently invasive science and technologies. In other words, there exist an intractable biopolitics and culture of food animals that consumers have found erstwhile surprisingly easy to accept and live with. We have taken pains to show that this 'surprisingly easy acceptance' does not come naturally and is, indeed, the outcome of the same disciplining governmentality exacted onto food animals (Heinz and Lee, 2009). The seduction of cheap meat blinds consumers to the real cost that undergirds their production. Besides the costs to the environment, health and workers' welfare, the well-being of food animals is especially obscured. This is despite various counter movements that broadly aim to reclaim food animals as lively beings and not reduce or normalise them into mere products.

There remains much to do. In this book we have written about what is broken in the livestock industry in broad strokes. The treatment of food animals and industry workers has to improve. Ever-increasing appetites for meat on the part of consumers need to be moderated too, for reasons of health and environment. Indeed, the very assumption we make of a normal diet needs to change. Perhaps what is urgently needed is research that can generate policy-oriented ideas to reform the meat industry, and not least to break the self-serving bonds between industry, institutions of science and regulators; research that complements the efforts of non-governmental organisations in their mission to reduce meat consumption; research that enables the smattering of companies involved in producing meat analogues; and research that can help narrow the cognitive dissonance between humans eating animals and humans' feeling for animals (Emel and Neo, 2015). The need for such research suggests that merely issuing guidelines to reduce the consumption of meat, as has been recently done by the Chinese government (Milman, 2016), is unlikely to cause material change in the demand for meat. The denormalisation of meat consumption is necessarily a tall order that, if even partially successful, will actualise a radically different world of meat, but it is surely a world we must strive towards.

References

Abbott, J., Campbell, L., Hay, C., Naesje, T., Ndumba, A. and Purvis, J., 2007. Rivers as resources, rivers as borders: community and transboundary management of fisheries in the Upper Zambezi River floodplains. *The Canadian Geographer*, 51(3), pp. 280–302.

Alexander, K.A., Blackburn, J.K., Vandewalle, M.E., Pesapane, R., Baipoledi, E.K. and Elzer, P.H., 2012. Buffalo, bush meat, and the zoonotic threat of brucellosis in Botswana. *PLoS One*, 7(3).

Allen, J. and Lavau, S., 2014. 'Just-in-time' disease: biosecurity, poultry and power. *Journal of Cultural Economy*, 8(3), pp. 342–360.

Alrøe, H.F., Vaarst, M. and Kristensen, E.S., 2001. Does organic farming face distinctive livestock welfare issues? A conceptual analysis. *Journal of Agricultural and Environmental Ethics*, 14(3), 275–299.

Althoff, P., 1979. Corporate control of agricultural commodity production, *Political Affairs*, 58(10), pp. 31–34.

American Veterinary Medical Association [AVMA], 2014. Welfare implications of tail docking of cattle. [online] 29 August. Available at: <https://www.avma.org/KB/Resources/LiteratureReviews/Pages/Welfare-Implications-of-Tail-Docking-of-Cattle.aspx> [Accessed 20 June 2016].

Amin, A., 1999. An institutionalist perspective on regional economic development, *International Journal of Urban and Regional Research*, 23, pp. 365–378.

Anderson, C.R. and McLachlan, S.M., 2012. Exiting, enduring and innovating: farm household adaptation to global zoonotic disease. *Global Environmental Change*, 22 (1), pp. 82–93.

Anderson, I., 2002. *Foot and mouth disease 2001: lessons to be learned inquiry report*. London: Stationery Office.

Anderson. K., 1997. A walk on the wild side: a critical geography of domestication. *Progress in Human Geography*, 21(4), pp. 463–485.

Andrews, J., 2013. The 10 biggest foodborne illness outbreaks of 2013. *Food Safety News*, [online] 27 December. Available at: <http://www.foodsafetynews.com/2013/12/the-10-biggest-u-s-outbreaks-of-2013/#.V2c2OKJHCoE> [Accessed 18 June 2016].

Armand-Lefevre, L., Ruimy, R. and Andremont, A., 2005. Clonal comparison of Staphylococcus aureus isolates from healthy pig farmers, human controls, and pigs. *Emerging Infectious Diseases*, 11(5), pp. 711–715.

Asano-Tamanoi, M., 1988. Farmers, industries, and the state: the culture of contract farming in Spain and Japan. *Comparative Studies in Society and History*, 30(3), pp. 432–452.

Asner, G.P., Elmore, A.J., Olander, L.P., Martin, R.E. and Harris, A.T., 2004. Grazing systems, ecosystem responses and global change. *Annual Review of Environment and Resources*, 29(1), pp. 261–299.

Atan, H. 2007. No more beta-agonist in pig feed. *New Straits Times*, 14 February.

Atkins, P.J., 1988. Redefining agricultural geography as the geography of food. *Area*, 20(3), pp. 281–283.

Atkins, P.J. and Bowler, I., 2001. *Food in society: economy, culture, geography*, London: Hodder Education.

Atkins, P.J. and Robinson, P.A., 2013. Coalition culls and zoonotic ontologies. *Environment and Planning A*, 45(6), pp. 1372–1386.

Atwood, M., 2003. *Oryx and Crake*. New York: Vintage.

Avieli, N., 2011. Dog meat politics in a Vietnamese town. *Ethnology*, 50(1), pp. 59–78.

Ayres, R.U., 1999. The second law, the fourth law, recycling and limits to growth. *Ecological Economics*, 29(3), pp. 473–483.

Azrul, M.K., 2015. Why I will never support hudud in Malaysia. *Malay Mail Online* [online] 19 March. Available at: <http://www.themalaymailonline.com/opinion/a zrul-mohd-khalib/article/why-i-will-never-support-hudud-in-malaysia> [Accessed 14 June 2016].

Babjee, A.M., Yap, T.C., Chee, Y.S., Candiah, P., Lim, E.S. and Cheong, Y.L., eds., 1983. *Proposal for the abatement of pollution from piggery waste in peninsular Malaysia*. Kuala Lumpur: Veterinary Division, Ministry of Agriculture.

Bager, F., Braggins, T.J., Devine, C.E., Graafhuis, A.E., Mellor, D.J.Tavener, A. and Upsdell, M.P., 1992. Onset of insensibility at slaughter in calves: effects of electroplectic seizure and exsanguination on spontaneous electrocortical activity and indices of cerebral metabolism. *Research in Veterinary Science*, 52(2), pp. 162–173.

Bailey, A.J., 2013. Migration, recession and an emerging transnational biopolitics across Europe. *Geoforum*, 44, pp. 202–210.

Bailey, R., Froggatt, A. and Wellesley, L., 2014. *Livestock – climate change's forgotten sector: global public opinion on meat and dairy consumption*. London: Chatham House Research Paper, Royal Institute of International Affairs. Available at https:// www.chathamhouse.org/sites/files/chathamhouse/field/field_document/20141203Lives tockClimateChangeBaileyFroggattWellesley.pdf [Accessed 20 June 2016].

Band, G. de Oliveira, Guimaraes, S.E.F., Lopes, P.S., de Oliveira Peixoto, J., Faria, D. A., Pires, A.V., de Castro Figueiredo, F., do Nascimento, C.S. and de Miranda Gomide, L.A., 2005. Relationship between the porcine stress syndrome gene and carcass and performance trains in F2 pigs resulting from divergent crosses. *Genetics and Molecular Biology*, 28(1), pp. 1415–1457.

Barclay, E., 2009. Meatless Mondays draw industry ire. *The Atlantic*, [online] 27 October. Available at: <http://www.theatlantic.com/health/archive/2009/10/meatless-mondays-draw-industry-ire/29092/> [Accessed 30 May 2014].

Barkema, A., Drabenstott, M. and Welch, K., 1991. The quiet revolution in the U.S. food market. *Economic Review*, [online] Available at: <https://www.kansascityfed. org/PUBLICAT/ECONREV/econrevarchive/1991/2q91bark.pdf> [Accessed 27 May 2014].

Barlett, D.L. and Steele, J.B., 2001. The empire of the pigs. *Time*, [online] 24 June. Available at: <http://content.time.com/time/magazine/article/0,9171,140572,00.html> [Accessed 30 May 2014].

Barton, J.R., 1997. Environment, sustainability and regulation in commercial aquaculture: the case of Chilean salmonid production. *Geoforum*, 28(3/4), pp. 313–328.

Bastian, B., Loughman, S., Haslam, N. and Radke, H.R.M., 2012. Don't mind meat? The denial of mind to animals used for human consumption. *Personality and Social Psychology Bulletin*, 38(2), pp. 247–256.

BBC, 2011. Australia bans all live cattle exports to Indonesia. [online] 8 June. Available at: <http://www.bbc.com/news/world-asia-pacific-13692211> [Accessed 3 March 2016].

BBC, 2013. Why are the British revolted by the idea of horsemeat? [online] 18 June. Available at: <http://www.bbc.com/news/magazine-21043368> [Accessed 2 June 2013].

Bell, C. and Neill, L., 2014. A vernacular food tradition and national identity in New Zealand. *Food, Culture and Society*, 17(1), pp. 49–64.

Bergeaud-Blackler, F., 2007. New challenges for Islamic ritual slaughter: a European perspective. *Journal of Ethnic and Migration Studies*, 33(6), pp. 965–980.

Bergsten, C., 2003. Causes, risk factors, and prevention of laminitis and related claw lesions. *Acta Veterinaria Scandinavica*, 44(Suppl 1), pp. S157–166.

Berita Harian, 2008. Total reared exceeds locals. *Berita Harian*, 11 April, p. 9.

Berkhout, N., 2010. Plans to provide spent hen renewable energy solution. *World Poultry*, [online] 23 February. Available at: <http://www.worldpoultry.net/Broilers/Health/2010/2/Plans-to-provide-spent-hen-renewable-energy-solution-WP007126W> [Accessed 7 October 2013].

Biermann, C. and Mansfield, B., 2014. Biodiversity, purity, and death: conservation biology as biopolitics. *Environment and Planning D: Society and Space*, 32(2), pp. 257–273.

Bjerklie, S., 2007. How 'new generation' meat plants forever changed New Zealand's industry. [online] May. <http://cemendocino.ucanr.edu/files/44389.pdf> [Accessed 6 August 2013].

Blackmore, D.K., 1984. Differences in behaviour between sheep and cattle during slaughter. *Research in Veterinary Science*, 37(2), pp. 223–226.

Blaikie, P., 1985. *The political economy of soil erosion in developing countries*. London: Longman.

Blaikie, P. and Brookfield, H., 1987. *Land degradation and society*. London: Methuen.

Bloomberg, 2013. Brin's $332,000 lab-grown burger has cake-like texture. [online] 6 August. Available at: <http://www.bloomberg.com/news/articles/2013-08-04/world-s-first-332-000-lab-grown-beef-burger-to-be-tasted> [Accessed 2 January 2014].

Bock, B. and Buller, H., 2013. Healthy, happy and humane: evidence in farm animal welfare policy. *Sociologia Ruralis*, 53(3), pp. 390–411.

Bokma-Bakker, M.H., Munnichs, G., Bracke, M.B.M., Visser, E.K., Schepers, F., Ursinus, W.W., Blokhuis, H.J., Gerritzen, M.A., Gast, E. Gaster, Evers, A.G., de Haan, M.H.A., van Mil, E.M., van Reenen, C.G. and Brom, F.W.A. 2009. *Animal-based welfare monitoring*. The Hague: Rathenau Institute.

Botreau, R., Veissier, I. and Perny, P., 2009. Overall assessment of animal welfare: strategy adopted in Welfare Quality®. *Animal Welfare*, 18(4), pp. 363–370.

Bourrie, M., 1999. Canada rejects bovine growth hormone. *Albion Monitor*, [online] 25 January. Available at: <http://www.albionmonitor.com/9901b/copyright/rbstcanada.html> [Accessed 7 January 2013].

Bowman, A., Mueller, K. and Smith, M., 2000. *Increased animal waste production from concentrated animal feeding operations (CAFOs): potential implications for public and environmental health*. Lincoln: Nebraska Center for Rural Health Research, Occasional Papers no. 2.

Boyd, W., 2001. Making meat: science, technology, and American poultry production. *Technology and Culture*, 42(4), pp. 631–664.

Brambell, F.W.R., 1965. *Report of the technical committee to enquire into the welfare of animals kept under intensive livestock husbandry systems* [Brambell Report]. London: Her Majesty's Stationery Office.

Braun, B. and Castree, N., eds., 1998. *Remaking reality: nature at the millennium*. London: Routledge.

Brooks, E., Emel, J., Robbins, P. and Jokisch, B., 2000. *The Ilano estacado of the U.S. southern High Plains: environmental transformation and the prospect for sustainability*. New York: United Nations University.

Brown, K. and Gilfoyle, D., 2010. *Healing the herds: disease, livestock economies, and the globalization of veterinary medicine*. Athens: Ohio University Press.

Brown, M. and Rasmussen, C., 2010. Bestiality and the queering of the human animal. *Environment and Planning D: Society and Space*, 28(1), pp. 158–177.

Bruce, A., 2011. Do sacred cows make the best hamburgers? *The Legal Regulation of Religious Slaughter of Animals*, 34(1), pp. 351–382.

Bryant, N., 2011. Did animal cruelty report lead to an over-reaction? *BBC*, [online] 8 June. Available at: <http://www.bbc.com/news/world-asia-pacific-13692280> [Accessed 5 May 2013].

Buchowski, M., 2003. Coming to terms with capitalism: an example of a rural community in Poland. *Dialectical Anthropology*, 27(1), pp. 47–68.

Buerkle, C.W., 2009. Metrosexuality can stuff it: beef consumption as (heteromasculine) fortification. *Text and Performance Quarterly*, 29(1), pp. 77–93.

Buller, H., 2013a. Individuation, the mass and farm animals. *Theory, Culture & Society*, 30(7/8), pp. 155–175.

Buller, H., 2013b. Animal geographies I. *Progress in Human Geography*, 38(2), pp. 308–318.

Buller, H. and Roe, E., 2013. Modifying and commodifying farm animal welfare: the economisation of the layer chickens. *Journal of Rural Studies*, 33, pp. 141–149.

Bureau of Labor Statistics [BLS], 2012/2013. *Occupational Outlook Handbook, 2012–13 Edition*, Slaughterers and Meat Packers, United States Department of Labor, [online]. Available at: <http://www.bls.gov/ooh/production/slaughterers-and-meat-packers.htm> [Accessed 30 July 2013].

Burgos, J.M., Ellington, B.A. and Varela, M.F., 2005. Presence of multidrug-resistant enteric bacteria in dairy farm topsoil. *Journal of Dairy Science*, 88(4), pp. 1391–1398.

Busch, L., 2008. Agricultural intensification and the environment. In: P.B. Thompson, ed., *The ethics of intensification: agricultural development and cultural change*. New York: Springer, pp. 149–157.

Bush, S.R., 2010. Governing 'spaces of interaction' for sustainable fisheries. *Tijdschrift voor Economische en Sociale Geografie*, 101(3), pp. 305–319.

Bush, S.R. and Duijf, M., 2011. Searching for (un)sustainabilty in Pangasius aquaculture: a political economy of quality in European retail. *Geoforum*, 42(2), pp. 185–196.

Butler, V., 2003. Inside the mind of a killer. *The Cyberactivist*, [online] 31 August. Available at: <http://cyberactivist.blogspot.com/2003/08/inside-mind-of-killer.html> [Accessed 29 July 2013].

Caple, I., McGown, P., Gregory, N. and Cusack, P., 2010. *Final report: Independent study into animal welfare conditions for cattle in Indonesia from point of arrival from Australia to slaughter*, Canberra: Australian Department of Agriculture, Fisheries and Forestry.

Caras, R.A., 2002. *A perfect harmony: the intertwining lives of animals and humans throughout history*, West Lafayette, IN: Purdue University Press.

Castree, N., 2005. *Nature*. London: Routledge.

Castree, N. and Braun, B., eds., 2001. *Social nature: theory, practice and politics*. Oxford: Blackwell.

Castree, N. and Nash, C., 2006. Posthuman geographies. *Social and Cultural Geography*, 7(4), pp. 501–504.

Chan, J.M., Stampfer, M.J., Giovannucci, E., Gann, P.H., Ma, J., Wilkinson, P., Hennekens, C.H. and Pollak, M., 1998. Plasma insulin-like growth factor-I and prostate cancer risk: a prospective study. *Science*, 279(5350), pp. 563–566.

Chang, K.H.K., 1978. Case of the dolphin kill. *Human Nature*, 1, pp. 84–87.

Chapple, A., Ziebland, S. and McPherson, A., 2004. Stigma, shame, and blame experienced by patients with lung cancer: qualitative study. *British Medical Journal*, 328(7454), pp. 1470–1474.

Chen, P.P., Wongsiri, S., Jamyanya, T., Rinderer, T.E., Vongsamanode, S., Matsuka, M., Sylvester, H.A. and Oldroyd, B.P., 1998. Honey bees and other edible insects used as human food in Thailand. *American Entomologist*, 44(1), pp. 24–29.

Cheong, S.-M., 2005. Korean fishing communities in transition: limitations of community-based resource management. *Environment and Planning A*, 37(7), pp. 1277–1290.

China Genebank, 2005. Yunnan Diannan small-ear pig protection scheme. [online] Available at: <http://www.genebank.cn/IMG/newspic/20058191562973499.htm> [Accessed 15 May 2008].

Cieślik, A., 2005. Regional characteristics and the location of foreign firms within Poland. *Applied Economics*, 37(8), pp. 863–874.

Clarke, N., 2008. From ethical consumerism to political consumption. *Geography Compass*, 2(6), pp. 1870–1884.

Cochrane, W.W., 1979. *The development of American agriculture: a historical analysis*. Minneapolis, MN: University of Minnesota Press.

Cockburn, A., 1996. A short, meat-oriented history of the world from Eden to the Mattole. *New Left Review*, 215, pp. 16–42.

Coff, C., 2006. *The taste for ethics: an ethic of food consumption*. Dordrecht: Springer.

Colombino, A. and Giaccaria, P., 2015. Breed contra beef: the making of Piedmontese cattle. In: J. Emel and H. Neo, eds., *Political ecologies of meat*. London: Routledge, pp. 161–177.

Compassion in World Farming [CIWF], 2013. Statistics: pigs. [online] 1 March. Available at: https://www.ciwf.org.uk/media/5235115/Statistics-pigs.pdf [Accessed 27 August 2014].

Compassion in World Farming Trust [CWFT], 2004. *The global benefits of eating less meat*. Petersfield: Compassion in World Farming Trust.

Conover, T., 2013. The way of all flesh: undercover in an industrial slaughterhouse. *Harper's Magazine*, May.

Counihan, C. and Van Esterik, P., eds., 2013. *Food and culture: a reader*. 3rd ed. London: Routledge.

Craddock, S., 1999. Embodying place: pathologizing Chinese and Chinatown in nineteenth-century San Francisco. *Antipode*, 31(4), pp. 351–371.

Crary, D., 2013. Pigs smart as dogs? Activists pose the question. *Associated Press*. [online] 29 July. Available at <http://news.yahoo.com/pigs-smart-dogs-activists-pose-073958905.html> [Accessed 1 August 2013].

Cumbers, A., MacKinnon, D. and McMaster, R., 2003. Institutions, power and space: assessing the limits to institutionalism in economic geography. *European Urban and Regional Studies*, 10(4), pp. 325–342.

Dadak, C., 2004. The case for foreign ownership in Poland. *Cato Journal*, 24(3), pp. 277–294.

Daley, G. and Sedgman, P., 2011. Australia bans live cattle exports to Indonesia. *Toronto Star* [online] 8 June. Available at: <https://www.thestar.com/business/economy/2011/06/08/australia_bans_live_cattle_exports_to_indonesia.html> [Accessed 10 April 2013].

Daly, C.C., Kallweit, E. and Ellendorf, F., 1988. Cortical function in cattle during slaughter: conventional captive bolt stunning followed by exsanguinations compared with shechita slaughter. *Veterinary Record*, 122(14), pp. 325–329.

Davis, K., 2009. *Prisoned chickens, poisoned eggs.* Summertown, TN: Book Publishing Company.

Dawkins, M.S., Donnelly, C.A. and Jones, T.A., 2004. Chicken welfare is influenced more by housing conditions than by stocking density. *Nature*, 427, pp. 342–344.

Dawley, S., 2007. Fluctuating rounds of inward investment in peripheral regions: semiconductors in the north east of England. *Economic Geography*, 83(1), pp. 51–73.

Dawson, M.D., Benson, E.R., Malone, G.W., Aphin, R.L., Estevez, I. and van Wicklen, G.L., 2006. Evaluation of foam-based mass depopulation methodology for floor-reared meat-type poultry operations. *Applied Engineering in Agriculture*, 22(5), pp. 787–794.

Dean, M., 2010. *Governmentality: power and rule in modern society.* 2nd ed. London: Sage.

Deutsch, S., 2005. Smithfield draws mixed reviews in Poland. *National Hog Farmer.* [online] 15 July. Available at: <http://nationalhogfarmer.com/mag/farming_smithfield_draws_mixed> [Accessed 4 September 2009].

Devendra, C., Mahendranathan, T., Wong, A. and Thamutaram, S., eds., 1972. *Symposium on the pig industry: proceedings of the symposium held at the State Veterinary Office, Selangor, 1–2 September 1971.* Kuala Lumpur: Ministry of Agriculture and Fisheries, Malaysia.

Dickson-Hoyle, S. and Reenberg, A., 2009. The shrinking globe: globalisation of food systems and the changing geographies of livestock production. *Danish Journal of Geography*, 109(1), pp. 105–112.

Digregorio, S., 2008. Test tube meat update: PETA 'civil war'. *Village Voice*, [online] 21 April. Available at: http://www.villagevoice.com/restaurants/test-tube-meat-update-peta-civil-war-6585869 [Accessed 10 April 2014].

Dillard, J., 2008. A slaughterhouse nightmare: psychological harm suffered by employees and the possibility of legal redress through legal reform. *Georgetown Journal on Poverty, Law and Policy.* 15(2), 391–409.

Doha News, 2012. Qatar: live sheep purchased from same shipment rejected by Bahrain free from disease. *Doha News*, [online] 5 September. Available at: <http://dohanews.co/post/30923044688/qatar-live-sheep-purchased-from-same-shipment-rejected> [Accessed 10 April 2013].

Dohoo, I.R., DesCôteaux, L., Leslie, K., Fredeen, A., Shewfelt, W., Preston, A. and Dowling, P., 2003. A meta-analysis review of the effects of recombinant bovine somatotropin: Effects on animal health, reproductive performance and culling. *Canadian Journal of Veterinarian Research*, 67(4), pp. 252–264.

Donner, H., 2008. New vegetarianism: food, gender and neoliberal regimes in Bengali middle-class families. *South Asia: Journal of South Asian Studies*, 31(1), pp. 143–169.

Dornisch, D., 2002. The evolution of post-socialist projects: trajectory shift and transitional capacity in a Polish region. *Regional Studies*, 36(3), pp. 307–321.

Douglas, M., 1966. *Purity and danger: an analysis of concepts of pollution and taboo.* London: Routledge.

Doyle, M., 2011. A bloody business. *Australian Broadcasting Corporation.* [podcast] 30 May. Available at: <http://www.abc.net.au/4corners/stories/2011/08/08/3288581.htm> [Accessed 10 April 2013].

DuPuis, E.M., 2002. *Nature's perfect food: how milk became America's drink.* New York: New York University Press.

Duval-Diop, D. and Grimes, J., 2005. Tales from two deltas: catfish fillets, high-value foods, and globalization. *Economic Geography,* 81(2), pp. 177–200.

Dzun, W., 1991. *Państwowe gospodarstwa rolne w rolnictwie polskim w latach 1944–1990.* Warsaw: Polska Akademia Nauk. Instytut Rozwoju Wsi i Rolnictwa.

Economist, 2014. Swine in China: Empire of the Pig. Available at: <http://www.economist.com/news/christmas-specials/21636507-chinas-insatiable-appetite-pork-symbol- countrys-rise-it-also> [Accessed 7 November 2016]

Edgar, J.L., Lowe, J.C., Paul, E.S. and Nicol, C.J., 2011. Avian maternal response to chick distress. *Proceedings of the Royal Society B, Biological Sciences,* 278(1721), pp. 3129–3134.

Eisnitz, G.A., 1997. *Slaughterhouse: the shocking story of greed, neglect, and inhumane treatment inside the US meat industry.* New York: Prometheus.

Elder, G., Wolch, J. and Emel, J., 1998. Le pratique sauvage: race, place, and the human–animal divide. In: J. Wolch and J. Emel, eds., *Animal geographies: place, politics, and identity in the nature-culture borderlands.* London: Verso, pp. 72–90.

Elferink, E.V., Nonhebel, S. and Moll, H.C., 2008. Feeding livestock food residue and the consequences for the environmental impact of meat. *Journal of Cleaner Production,* 16(12), pp. 1227–1233.

Emel, J., 1998. Are you man enough, big and bad enough? Wolf eradication in the US. In: J. Wolch and J. Emel, eds., *Animal geographies: place, politics, and identity in the nature-culture borderlands.* London: Verso, pp. 91–116.

Emel, J. and Hawkins, R., 2010. Is it really easier to imagine the end of the world than the end of industrial meat? *Human Geography,* 3(2), pp. 35–48.

Emel, J. and Neo, H., 2011. Killing for profit: global livestock industries and their socio-ecological implications. In: R. Peet, P. Robbins and M.J. Watts, eds., *Global political ecology.* London: Routledge, pp. 67–83.

Emel, J. and Neo, H., 2015. Conclusion: affect and attribution. In: J. Emel and H. Neo, eds., *Political ecologies of meat.* London: Routledge, pp. 21–41.

Enticott, G., 2008. The spaces of biosecurity: prescribing and negotiating solutions to bovine tuberculosis. *Environment and Planning A,* 40(7), pp. 1568–1582.

Enticott, G., 2014. Relational distance, neoliberalism and the regulation of animal health. *Geoforum,* 52(1), pp. 42–50.

Esposito, R., 2008. *Bíos: biopolitics and philosophy.* Minneapolis: University of Minnesota Press.

Ettlinger, N., 2011. Governmentality as epistemology. *Annals of the Association of American Geographers,* 101(3), pp. 537–560.

European Commission, 2010. *Report from the Commission to the European Parliament and the Council on animal cloning for food production.* COM (2010) 585, 19 October. Brussels: European Commission.

European Commission, 2014. Horse meat: one year after – actions announced and delivered! [online]. Available at: <http://ec.europa.eu/food/food/horsemeat/> [Accessed 2 May 2014].

Eurostat, 2010. Europe in figures. Available at: <http://ec.europa.eu/eurostat/web/p roducts-statistical-books/-/KS-CD-10-220> [Accessed 7 November 2016]

Ewell, P.R., 1963. *Contract farming, U.S.A.* Danville, IL: Interstate Printers and Publishers.

Eyes on Animals, 2013. Importance of access during transport. [online] February. Available at: <http://www.eyesonanimals.com/resources/special-reports/> [Accessed 20 June 2016].

Fa, J.E., Currie, D. and Meeuwig, J., 2003. Bushmeat and food security in the Congo Basin: linkages between wildlife and people's future. *Environmental Conservation*, 1(1), pp. 71–78.

Fa, J.E., Peres, C.A. and Meeuwig, J., 2002. Bushmeat exploitation in tropical forests: an intercontinental comparison. *Conservation Biology*, 16(1), pp. 232–237.

Fagan, R., 1997. Local food/global food: globalization and local restructuring. In: R. Lee and J. Wills, eds., *Geographies of economies*. London: Arnold, pp. 197–208.

Farm Animal Welfare Council [FAWC], 2009. Farm animal welfare in Great Britain: past, present and future. London: FAWC. [online] Available at: https://www.gov.uk/ government/uploads/system/uploads/attachment_data/file/319292/Farm_Animal_ Welfare_in_Great_Britain_-_Past__Present_and_Future.pdf [Accessed 14 May 2016].

Ferguson, S., 2011. Interview with Dr Temple Grandin. *Australian Broadcasting Corporation*, Transcript [online] 30 May.Available at: <http://www.abc.net.au/4corners/ content/2011/s3230885.htm> [Accessed 10 November 2013].

Ferguson, S. and Masters, D., 2012. Another bloody business. *Australian Broadcasting Corporation* [online] 7 November. Available at: <http://www.abc.net.au/4corners/ stories/2012/11/02/3623727.htm> [Accessed 10 November 2013].

Fiala, N., 2008. Meeting the demand: an estimation of potential greenhouse gas emissions from meat production. *Ecological Economics*, 67(3), pp. 412–419.

Fielding, R., 2013. Whaling futures: a survey of Faroese and Vincentian youth on the topic of artisanal whaling. *Society and Natural Resources*, 26(7), pp. 810–826.

Fine, B., 1994a. Towards a political economy of food. *Review of International Political Economy*, 1(3), pp. 519–545.

Fine, B., 1994b. A response to my critics. *Review of International Political Economy*, 1(3), pp. 579–586.

Fine, B., 1998. *The political economy of diet, health and food policy.* London: Routledge.

Fitzgerald, A.J., Kalof, L. and Dietz, T., 2009. Slaughterhouses and increased crime rates: an empirical analysis of the spillover from 'the Jungle' into the surrounding community. *Organization & Environment*, 22(2), pp. 158–184.

Foer, J.S., 2009. *Eating animals.* New York: Little, Brown.

Food and Agriculture Organization [FAO], 2002. Manual on the diagnosis of Nipah virus infection in animals. Bangkok: Food and Agriculture Organization of the United Nations, Regional Office for Asia and the Pacific. [online] Available at: <http://www.fao.org/docrep/005/ac449e/ac449e00.htm> [Accessed 24 May 2013].

Food and Agriculture Organization [FAO], 2005. The globalizing livestock sector: impact of changing markets. Committee on Agriculture. Rome: Food and Agricultural Organization of the United Nations, [online] 13–16 April. Available at: <http://www.fao.org/docrep/meeting/009/j4196e.htm> [Accessed 10 May 2011].

Food and Agriculture Organization [FAO], 2006. Livestock a major threat to environment. *Newsroom*, [online] 29 November. Available at: <http://www.fao.org/news room/en/news/2006/1000448/ > [Accessed 7 Dec 2006].

Food and Agriculture Organization [FAO], 2011. *Global food losses and food waste: extent, causes and prevention.* Rome: Food and Agricultural Organization of the United Nations.

Food and Water Watch, 2010. *Factory farm nation: How America turned its livestock farms into factories.* Washington, DC: Food & Water Watch.

Foote, R.H., 2002. The history of artificial insemination: notations and notables. *Journal of Animal Science,* 80, pp. 1–10.

Foucault, M., 1979. *Discipline and punish: the birth of the prison.* New York: Vintage.

Foucault, M., 1990. *The history of sexuality: an introduction,* vol. 1. New York: Knopf/Doubleday.

Foucault, M., 2003. *Abnormal: lectures at the Collège de France 1974–1975.* London: Verso.

Foucault, M., 2009. *Security, territory, population: lectures at the Collège de France 1977–1978.* New York: Palgrave Macmillan.

Fowler, K., 2013. Slaughterhouse secrets. *Animal Aid* [online] 22 February. Available at: <http://www.animalaid.org.uk/h/n/ABOUT/blog//2833//> [Accessed 20 June 2016].

Fox, N. and Ward, K., 2007. Health, ethics and environment: a qualitative study of vegetarian motivations. *Appetite,* 50(2/3), pp. 422–429.

Fox, R., 2006. Animal behaviours, post-human lives: everyday negotiations of the animal–human divide in pet-keeping. *Social & Cultural Geography,* 7(4), pp. 525–537.

Frank, J., 2007. Meat as a bad habit: a case for positive feedback in consumption preferences leading to lock-in. *Review of Social Economy,* 65(3), pp. 319–348.

Fraser, D., 2009. Assessing animal welfare: different philosophies, different scientific approaches. *Zoo Biology,* 28(6), pp. 507–518.

Freese, B., 2011. Pork powerhouse: Midwest sows on the rise. [online] Available at: <http://www.agriculture.com/livestock/hogs/pk-powerhouses-midwest-sows-on-rise_283-ar19461> [Accessed 7 April 2013].

Friedman, H., 1993. The political economy of food: a global crisis article. *New Left Review,* 197, pp. 29–573.

Fuglie, K., Heisey, P., King, J.L., Pray, C.E., Day-Rubenstein, K., Schimmelpfennig, D., Wang, S.L. and Karmarkar-Deshmukh, R., 2011. Research investments and market structure in the food processing, agricultural input, and biofuel industries worldwide. Washington: United States Department of Agriculture, Economic Research Report no. ERR-130.

Gerritzen, M.A., Lambooij, B., Reimert, H., Stegeman, A. and Spruijt, B., 2004. On-farm euthanasia of broiler chickens: effects of different gas mixtures on behavior and brain activity. *Poultry Science,* 83(8), pp. 1294–1301.

Gisolfi, M.R., 2006. From crop lien to contract farming: the roots of agribusiness in the American South, 1929–1939. *Agricultural History,* 80(2), pp. 167–189.

Golab, G.C., 2007. Testimony concerning a review of the welfare of animals in agriculture. Hearing before the Subcommittee on Livestock, Dairy, and Poultry Committee on Agriculture, US House of Representatives, 110th Congress, First Session, 8 May.

Goldschmidt, W., 1998. Conclusion: The urbanization of rural America. In: K.M. Thu and E.P. Durrenberger, eds., *Pigs, profits, and rural communities,* Albany: State University of New York Press, pp. 183–198.

Gomez, E.T. and Saravanamuttu, J., eds., 2012. *The New Economic Policy in Malaysia: affirmative action, ethnic inequalities and social justice.* Singapore: NUS Press.

Goodman, D., 2002. Rethinking food production-consumption: integrative perspectives. *Sociologia Ruralis,* 42(4), pp. 271–277.

Goodman, D. and Redclift, M., 1994. Constructing a political economy of food. *Review of International Political Economy*, 1(3), pp. 547–552.

Goodman, M.K., Maye, D. and Holloway, L., 2010. Ethical foodscapes? Premises, promises, and possibilities. *Environment and Planning A*, 42(8), pp. 1782–1796.

Gouveia, L. and Juska, A., 2002. Taming nature, taming workers: constructing the separation between meat consumption and meat production in the U.S. *Sociologia Ruralis*, 42(4), pp. 370–390.

GRAIN, 2008. Making a killing from hunger. [online] 28 April. Available at: http://www.grain.org/articles/?id=39 [Accessed 27 November 2013].

Grandin, T., 2013. Electric stunning of pigs and sheep. [online] Available at: <http://www.grandin.com/humane/elec.stun.html> [Accessed 28 May 2014].

Grandin, T. and Regenstein, J.M., 1994. Religious slaughter and animal welfare: a discussion for meat scientists. *Meat Focus International*, March: 115–123.

Gray, G.C., McCarthy, T., Capuano, A.W., Setterquist, S.F., Olsen, C.W., Alavanja, M.C. and Lynch, C.F., 2007. Swine workers and swine influenza virus infections. *Emerging Infectious Diseases*, 13(12), pp. 1871–1878.

Greenpeace, 2009. Slaughtering the Amazon. [online] 1 June, Available at: <http://www.greenpeace.org/international/press/reports/slaughtering-the-amazon> [Accessed 28 May 2014].

Gregory, N.G. and Shaw, F., 2000. Penetrating captive bolt stunning and exsanguination of cattle in abattoirs. *Journal of Applied Animal Welfare Science*, 3(3), pp. 215–230.

Gregory, N.G. and Wotton, S.B., 1984. Time to loss of brain responsiveness following exsanguination in calves. *Research in Veterinary Science*, 37(2), pp. 141–143.

Gregory, N.G., von Wenzlawowicz, M. and von Holleben, K., 2009. Blood in the respiratory tract during slaughter with and without stunning in cattle. *Meat Science*, 82(1), pp. 13–16.

Gunderson, R., 2011. From cattle to capital: exchange value, animal commodification, and barbarism. *Critical Sociology*, 39(2), pp. 259–275.

Guthman, J., 1998. Regulating meaning, appropriating nature: the codification of California organic agriculture. *Antipode*, 30(2), pp. 135–154.

Guthman, J., 2003. 'Fast food/organic food: reflexive tastes and the making of 'yuppie chow'. *Social and Cultural Geography*, 4(1), pp. 45–58.

Guthman, J., 2011. *Weighing in: obesity, food justice, and the limits of capitalism.* Berkeley: University of California Press.

Hagen, C.A., Pitman, J.C., Loughin, T.L., Sandercock, B.K., Robel, R.J. and Applegate, R.D., 2011. Potential impacts of anthropogenic features on lesser prairie-chicken habitat use. *Studies in Avian Biology*, 39, pp. 63–75.

Hagen, K., van den Bos, R. and de Cock Buning, T., 2011. Editorial: Concepts of animal welfare. *Acta Biotheoretica*, 59(2), pp. 93–103.

Hall, D., 2010. Food with a visible face: traceability and the public promotion of private governance in the Japanese food system. *Geoforum*, 41(5), pp. 826–835.

Hall, D.C., Ehui, S. and Delgado, C., 2004. The livestock revolution, food safety, and small-scale farmers: why they matter to us all. *Journal of Agricultural and Environmental Ethics*, 17(4/5), pp. 425–444.

Halverson, N., 2014. How will China feed its middle class? Available at: <http://www.pbs.org/newshour/bb/china-porkcir/> [Accessed 7 November 2016]

Halweil, B. and Nierenberg, D., 2008. Meat and seafood: the most costly ingredients in the global diet. In: The Worldwatch Institute, ed., *State of the world 2008: innovations for a sustainable economy*. Washington, DC: Worldwatch Institute, pp. 61–74.

Hamidah, A., 2007. No more beta-agonist in pig feed. *New Straits Times*, 14 February.

Hardy, J., 2006. Bending workplace institutions in transforming economies: foreign investment in Poland. *Review of International Political Economy*, 13(1), pp. 129–151.

Harrison, C.M., Flynn, A. and Marsden, T., 1997. Contested regulatory practice and the implementation of food policy: exploring the local and national interface. *Transactions of the Institute of British Geographers*, 22(4), pp. 473–487.

Harrison, R., 1964. *Animal machines*. London: Vincent Stuart.

Hartwick, E., 1998. Geographies of consumption: a commodity chain approach. *Environment and Planning D*, 16(4), pp. 423–437.

Hauser, C., 2002. Work at slaughterhouse is halted after graphic undercover videos. *New York Times* [online], 22 August. Available at: <http://thelede.blogs.nytimes.com/2012/08/22/work-at-slaughterhouse-is-halted-after-graphic-undercover-videos/> [Accessed 29 July 2013].

Hawthorne, M., 2005. From fiction to fork. *Satya*. [online] September. Available at: <http://www.satyamag.com/sept05/hawthorne.html> [Accessed 29 July 2013].

Haynes, R.P., 2011. Competing conceptions of animal welfare and their ethical implications for the treatment of non-human animals. *Acta Biotheoretica*, 59(2), pp. 105–120.

Healey, P., 2006. Transforming governance: challenges of institutional adaptation and a new politics of space. *European Planning Studies*, 14(3), pp. 299–320.

Heederik, D., Sigsgaard, T., Thorne, P.S., Kline, J.N., Avery, R., Bønløkke, J.H., Chrischilles, E.A., Dosman, J.A., Duchaine, C., Kirkhorn, S.R., Kulhankova, K. and Merchant, J.A., 2007. Health effects of airborne exposures from concentrated animal feeding operations. *Environmental Health Perspectives*, 115(2), pp. 298–302.

Heffernan, W.D., 1972. Sociological dimensions of agricultural structures in the United States. *Sociologia Ruralis*, 12(2), pp. 481–499.

Heinrich Böll Foundation, 2014. *Meat atlas: facts and figures about the animals we eat*. Berlin: Heinrich Böll Stiftung and Friends of the Earth Europe. [online] <http://www.boell.de/en/2014/01/07/meat-atlas> [Accessed 2 March 2014].

Heinz, B. and Lee, R., 2009. Getting down to the meat: the symbolic construction of meat consumption. *Communication Studies*, 49(1), pp. 86–99.

Hendrickson, M., and Heffernan, W., 2007. Concentration of agricultural markets. [online] April. Available at: <http://www.foodcircles.missouri.edu/07contable.pdf> [Accessed 1 September 2012].

Hercock, M. and Tonts, M., 2004. From the rangelands to the Ritz: geographies of kangaroo management and trade. *Geography*, 89(3), pp. 214–225.

Hernandez, V., 2012. Pakistan accepts 200,000 stranded Australian sheep rejected by Bahrain. *International Business Times* [online] 5 September. Available at: <http://www.ibtimes.com.au/pakistan-accepts-200000-stranded-australian-sheep-rejected-bahrain-1300286> [Accessed 10 April 2013].

Hilliard, S.B., 1969. Pork in the antebellum South: the geography of self-sufficiency. *Annals of the Association of American Geographers*, 59(3), pp. 461–480.

Hinchliffe, S. and Ward, K.J., 2014. Geographies of folded life: how immunity reframes biosecurity. *Geoforum*, 53, pp. 136–144.

Hoff, S.J., Hornbuckle, K.C., Thorne, P.S., Bundy, D.S., and O'Shaughnessy, P.T., 2002. Emissions and community exposures from CAFOs. In: Iowa State University and the University of Iowa Study Group, *Iowa concentrated animal feeding operations air quality Study*, pp. 45–85.

Holloway, L., Kneafsey, M., Venn, L., Cox, R., Dowler, E. and Tuomainen, H., 2007. Possible food economies: a methodological framework for exploring food production-consumption relationships. *Sociologia Ruralis*, 47(1), pp. 1–19.

Holmes, M.D., Pollak, M.N., Willett, W.C. and Hankinson, S.E., 2002. Dietary correlates of plasma insulin-like growth factor I and insulin-like growth factor binding protein 3 concentrations. *Cancer Epidemiology, Biomarkers, and Prevention*, 11(9), pp. 852–861.

Hopkins, P.D. and Dacey, A., 2008. Vegetarian meat: could technology save animals and satisfy meat eaters? *Journal of Agricultural and Environmental Ethics*, 21(6), pp. 579–596.

Hopkins, R.F. and Puchala, D.J., eds., 1978. *The global political economy of food*. Madison: University of Wisconsin Press.

Horrigan, L., Lawrence, R.S. and Walker, P., 2007. How sustainable agriculture can address the environmental and human health harms of industrial agriculture. *Environmental Health Perspectives*, 110(5), pp. 445–457.

Horwitz, H., 2001. Uses of racialism: hybrids, race and cultural legibility. *Western Humanities Review*, 55(1), pp. 43–64.

Hovorka, A.J., 2006. The No. 1 Ladies' Poultry Farm: a feminist political ecology of urban agriculture in Botswana. *Gender, Place and Culture*, 13(3), pp. 207–225.

Hribar, C., 2010. *Understanding concentrated animal feeding operations and their impact on communities*. Bowling Green, OH: National Association of Local Boards of Health. [online] Available at: <http://www.cdc.gov/nceh/ehs/docs/understanding_cafos_nalboh.pdf> [Accessed 26 July 2013].

Huang, H. and Miller, G.Y., 2006. Citizen complaints, regulatory violations, and their implications for swine operations in Illinois. *Review of Agricultural Economics*, 28(1), pp. 89–110.

Huber, M.T., 2013. *Lifeblood: oil, freedom, and the forces of capital*. Minneapolis: University of Minnesota Press.

Hudson, P., 1983. Proto-industrialization: the case of West Riding wool textile industry in the eighteenth and early nineteenth century. In: M. Berg and P. Hudson, eds., *Manufacture in town and country before the factory*. Cambridge: Cambridge University Press, pp. 124–144.

Human Rights Watch, 2004. *Blood, sweat, and fear: workers' rights in U.S. meat and poultry plants*. New York: Human Rights Watch.

Humane Society of the United States [HSUS], 2008. Rampant animal cruelty at California slaughter plant: Undercover investigation finds abuses at major beef supplier to America's school lunch program. [video] Available at: <http://www.humanesociety.org/news/news/2008/01/undercover_investigation_013008.html> [Accessed 1 June 2014].

Hussein, M. and Lim, C.B., 1972. Future potential and prospects of the pig industry. In: C. Devendra, T. Mahendranathan, A. Wong and S. Thamutaram, eds., *Symposium on the pig industry: proceedings of the symposium held at the State Veterinary Office, Selangor, 1–2 September 1971*. Kuala Lumpur: Ministry of Agriculture and Fisheries, Malaysia.

Huxley, M., 2007. Geographies of governmentality. In: J.W. Crampton and S. Elden, eds., *Place, knowledge and power: Foucault and geography*. Aldershot: Ashgate, pp. 163–167.

Idris, S.M.M., 2006. Banned substance readily available. *New Straits Times*, 10 May.

Iggulden, T., 2011. Indonesia protests live cattle export ban. *Lateline*, Australian Broadcasting Corporation, Interview transcript [online] 8 June. Available at: <http://www.abc.net.au/lateline/content/2011/s3239274.htm> [Accessed 10 November 2013].

Ikerd, J., 2003. Corporate livestock production: implications for rural North America. In: A.M. Ervin, C. Holtslander, D. Qualman and R. Sawa, eds., *Beyond factory farming: corporate hog barns and the threat to public health, the environment, and Rural Communities.* Saskatoon: Canadian Centre for Policy Alternatives – Saskatchewan, pp. 29–38.

Ilbery, B. and Maye, D., 2005. Alternative (shorter) food supply chains and specialist livestock products in the Scottish–English borders. *Environment and Planning A,* 37(5), pp. 823–844.

Ilea, R.C., 2009. Intensive livestock farming: global trends, increased environmental concerns, and ethical solutions. *Journal of Agriculture and Environmental Ethics,* 22(2), pp. 153–167.

International Vegetarian Union, 2013. About IVU. [online] Available at: <http://www.ivu.org/index.php?option=com_content&view=article&id=315&Itemid=268> [Accessed 20 June 2016].

Islam, Q.M.Q., 2007. *The Islamic concept of animal slaughter.* Kuala Lumpur: A.S. Noordeen.

Jabs, J., Sobal, J. and Devine, C.M., 2000. Managing vegetarianism: identities, norms and interactions. *Ecology of Food and Nutrition,* 39(5), pp. 375–394.

Jackson, P., ed. 1987. *Race and racism: essays in social geography.* London: Allen and Unwin.

Jarosz, L., 2009. Energy, climate change, meat and markets: mapping the coordinates of the current world food crisis. *Geography Compass,* 3(6), pp. 2065–2083.

Jensen, P., 1986. Observations on the maternal behavior of free-ranging domestic pigs. *Applied Animal Behaviour Science,* 16(2), pp. 131–142.

Jessop, B., 2001. Institutional re(turns) and the strategic-relational approach. *Environment and Planning A,* 33(7), pp. 1213–1235.

Jin, S.K., Hur, I.C., Jeong, J.Y., Choi, Y.J., Choi, B.D., Kim, B.G. and Hur, S.J., 2011. The development of imitation crab sticks by substituting spent laying hen meat for Alaska Pollack. *Poultry Science,* 90(8), pp. 1799–1808.

Johnson, D.G., 1978. World food institutions: a 'liberal' view. In: R.F. Hopkins and D.J. Puchala, eds., *The global political economy of food.* Madison: University of Wisconsin Press, pp. 65–82.

Jones, T., 2011. MLA knew of 'issues' in Indonesia abattoirs. *Australian Broadcasting Corporation,* Interview transcript [online] 8 June. Available at: <http://www.abc.net.au/lateline/content/2011/s3239281.htm> [Accessed 10 April 2013].

Kalland, A., 1992. Whose whale is that? Diverting the commodity path. In: M.M.R. Freeman and U. Kreuter, eds., *Elephants and whales: resources for whom.* Basel: Gordon and Breach, pp. 159–186.

Kalof, L., Dietz, T., Stern, P.C. and Guagnano, G.A., 1999. Social psychological and structural influences on vegetarian beliefs. *Rural Sociology,* 64(3), pp. 500–511.

Kandel, W. and Parrado, E.A., 2005. Restructuring of the US meat processing industry and new Hispanic migrant destinations. *Population and Development Review,* 31(3), pp. 447–471.

Keeney, R., 2008. Community impacts of CAFOs: labor markets. Purdue University Extension ID 362. [online] Available at: <https://www.extension.purdue.edu/extmedia/ID/ID-362-W.pdf> [Accessed 30 May 2014].

Kloppenburg, J.R., 2004. *First the seed: the political economy of plant biotechnology.* 2nd ed. Madison: University of Wisconsin Press.

Kolpin, D.W., Furlong, E.T., Meyer, M.T., Thurman, E.M., Zaugg, S.D., Barber, L.B. and Buxton, H.T., 2002. Pharmaceuticals, hormones, and other organic wastewater contaminants in U.S. streams, 1999–2000: a national reconnaissance. *Environmental Science and Technology*, 36(6), pp. 1202–1211.

Kretzer, M., 2012. Video shows pigs mutilated, beaten, duct-taped. *People for the Ethical Treatment of Animals.* [online] 1 February.Available online: <http://www.peta.org/blog/pigs-mutilated-beaten-duct-taped/> [Accessed 30 May 2014].

Kristensen, H.H., Berry, P.S. and Tinker, D.B., 2001. Depopulation systems for spent hens – a preliminary evaluation in the United Kingdom. *Journal of Applied Poultry Research*, 10(2), pp. 172–177.

Kümpel, N.F., Milner-Gulland, E.J., Cowlishaw, G. and Rowcliffe, J.M., 2010. Incentives for hunting: the role of bushmeat in the household economy in rural Equatorial Guinea. *Human Ecology*, 38(2), pp. 251–264.

Labben, M., 2014. Against value: accumulation in the oil industry and the biopolitics of labour under finance. *Antipode*, 46(2), pp. 477–496.

Labrianidis, L., 1995. Flexibility in production through subcontracting: the case of the poultry meat industry in Greece. *Environment and Planning A*, 27(2), pp. 193–209.

Lang, T. and Barling, D., 2012. Food security and food sustainability: reformulating the debate. *Geographical Journal*, 178(4), pp. 313–326.

Langdon, W.C., 1945. The Argentine meat question. *Geographical Review*, 35(4), pp. 634–646.

Latour, B., 2004. *Politics of nature: how to bring the sciences into democracy.* Cambridge, MA: Harvard University Press.

Leahy, E., Lyons, S. and Tol, R., 2010. *An estimate of the number of vegetarians in the world.* Working Paper no. 340. Dublin: Economic and Social Research Institute.

Leitzmann, C., 2003. Nutrition ecology: the contribution of vegetarian diets. *American Journal of Clinical Nutrition*, 78(3), pp. 6575–6595.

Lévi-Strauss, C., 1963. *Totemism*, trans. R. Needham. Boston: Beacon Press.

Liebers, V., Raulf-Heimsoth, M. and Brüning, T., 2008. Health effects due to endotoxin inhalation. *Archives of Toxicology*, 82(4), pp. 203–210.

Lim, G. and Neo, H., 2014. The economic geographies of aquaculture. *Geography Compass*, 8(9), pp. 665–676.

Little, P.D. and Watts, M.J., 1994. Introduction. In: P.D. Little and M.J. Watts, eds., *Living under contract: contract farming and agrarian transformation in sub-Saharan Africa.* Wisconsin: University of Wisconsin Press, pp. 3–18.

LiveCorp, 2013. An introduction. [online] Available at: <http://www.livecorp.com.au/about-us/introduction> [Accessed 20 June 2016].

Long, N. and van der Ploeg, J.D., 1988. New challenges in the sociology of rural development: a rejoinder to Peter Vandergeest. *Sociologia Ruralis*, 28(1), pp. 30–41.

Lowe, P., Marsden, T. and Whatmore, S., eds., 1994. *Regulating agriculture.* London: Fulton.

Lumb, S., 2010. Danes aiming for 35 piglets per sow per year. *Pig Progress* [online] Available at: <http://www.pigprogress.net/Home/General/2010/12/Danes-aiming-for-35-piglets-per-sow-per-year-PP007030W/> [Accessed 28 May 2014].

Lusk, J.L., 2013. The role of technology in the global economic importance and viability of animal protein production. *Animal Frontiers*, 3(3), pp. 20–27.

MacDonald, J. and McBride, W., 2009. *The transformation of U.S. livestock agriculture: scale, efficiency, and risks.* United States Department of Agriculture, Economic Information Bulletin no. 43. [online] Available at: <http://www.ers.usda.gov/publica tions/eib-economic-information-bulletin/eib43.aspx#.U3-aFChhvqw> [Accessed 23 May 2014].

MacLachlan, I., 2005. Feedlot growth in southern Alberta: a neo-Fordist interpretation. In: S.J. Essex, A.W. Gilg and R.B. Yarwood, eds., *Rural change and sustainability: agriculture, the environment and communities.* London: CABI Publishing, pp. 28–47.

MacLachlan, I., 2015. Evolution of a revolution: meat consumption and livestock production in the developing world. In: J. Emel and H. Neo, eds., *Political ecologies of meat.* London: Routledge, pp. 21–41.

MacLeod, G., 2001. Beyond soft institutionalism: accumulation, regulation, and their geographical fixes. *Environment and Planning A,* 33(7), pp. 1145–1167.

Mann, S. and Kogl, H., 2003. On the acceptance of animal production in rural communities. *Land Use Policy,* 20(3), pp. 243–252.

Mansfield, B., 2003. From catfish to organic fish: making distinctions about nature as cultural economic practice. *Geoforum,* 34(3), pp. 329–342.

Mansfield, B., 2006. Assessing market-based environmental policy using a case study of North Pacific fisheries. *Global Environmental Change,* 16(1), pp. 29–39.

Mansfield, B., 2011. Is fish health food or poison? Farmed fish and the material production of un/healthy nature. *Antipode,* 43(2), pp. 413–434.

Markey, S., 2005. Building local development institutions in the hinterland: a regulationist perspective from British Columbia, Canada. *International Journal of Urban and Regional Research,* 29(2), pp. 358–374.

Marsden, T., Munton, R., Ward, N. and Whatmore, S., 1996. Agricultural geography and the political economy approach: a review. *Economic Geography,* 72(4), pp. 361–375.

Martin, R., 2000. Institutional approaches in economic geography. In: E. Sheppard and T.J. Barnes, eds., *A companion to economic geography.* Oxford: Blackwell, pp. 77–94.

Marx, K. [1867] 1977. *Capital: a critique of political economy,* vol. 1. New York: Vintage.

Maurer, D., 2002. *Vegetarianism: movement or moment?* Philadelphia: Temple University Press.

McAlpine, C.A., Syktus, J., Deo, R.C., Lawrence, P.J., McGowan, H.A., Watterson, I. G. and Phinn, S.R., 2007. Modeling the impact of historical land cover change on Australia's regional climate. *Geophysical Research Letters,* 34(22).

McClaskey, J., 2004. Economic and cost considerations. In: *Carcass disposal: a comprehensive review.* Report prepared by the National Agricultural Biosecurity Center Consortium, United States Department of Agriculture, pp. 40–42. [online] Available at: <http://amarillo.tamu.edu/files/2011/01/draftreport.pdf> [Accessed 28 May 2014].

McCosker, L., 2012. *Free range pig farming: starting out in pastured pigs.* n.p.

McMichael, P., 1992. Tensions between national and international control of the world food order: contours of a new food regime. *Sociological Perspectives,* 35(2), pp. 343–365.

McMichael, P., ed. 1994. *The global restructuring of agro-food systems.* Ithaca: Cornell University Press.

McMichael, P., 2009. A food regime genealogy. *Journal of Peasant Studies,* 36(1), pp. 139–169.

McMurtry, M.J., Wales, D.L., Scheider, W.A., Beggs, G.L. and Dimond, P.E., 1989. Relationship of mercury concentrations in lake trout (*Salvelinus namaycush*) and smallmouth bass (*Micropterus dolomieui*) to the physical and chemical characteristics of Ontario lakes. *Canadian Journal of Fish and Aquatic Science* 46(3), pp. 426–434.

Meat and Livestock Australia [MLA], 2013. Improving standing stun restraint in Indonesia. Final Report by Meat and Livestock Australia. [online] Available at: <www.livecorp.com.au/LC/files/c9/c9f0cc25-59ec-4cac-82ab-07fd0159c645.pdf> [Accessed 20 June 2016].

Meat Free Monday, 2014. Meat Free Monday celebrates five years. [online] Available at: http://www.meatfreemondays.com/meat-free-monday-celebrates-five-years/ [Accessed 18 June 2014].

Meersman, T., 2001. Olivia area hog operation continues to violate air rules. *Minneapolis Star Tribune*, 14 June.

Merchant, J.A., Naleway, A.L., Svendsen, E.R., Kelly, K.M., Burmeister, L.F., Stromquist, A.M., Taylor, C.D., Thorne, P.S., Reynolds, S.J., Sanderson, W.T. and Chrischilles, E.A., 2005. Asthma and farm exposures in a cohort of rural Iowa children. *Environmental Health Perspectives*, 113(3), pp. 350–356.

Micek, G., Neo, H. and Górecki, J., 2011. Foreign direct investment, institutional context and the changing Polish pig industry. *Geografiska Annaler: Series B, Human Geography*, 93(1), pp. 41–55.

Miller, I., 2011. Evangelicalism and the early vegetarian movement in Britain c.1847–1860. *Journal of Religious History*, 35(2), pp. 199–210.

Milman, O., 2016. China's plan to cut meat consumption by 50% cheered by climate campaigners. *The Guardian*, [online] 20 June. Available at: <https://www.theguardia n.com/world/2016/jun/20/chinas-meat-consumption-climate-change> [Accessed 22 June 2016].

Milner-Gulland, E.J., Bennett, E.L. and the SCB, 2003. Annual Meeting Wild Meat Group [WMG], 2003. Wild meat: the bigger picture. *Trends in Ecology and Evolution*, 18(7), pp. 351–357.

Ministry of Agriculture, Malaysia, 1965. Proposals for the development of the livestock industry and the reorganization of the veterinary services: report of meeting on pig production. In: *Proposals for the development of the livestock industry and the reorganization of the veterinary services*. Kuala Lumpur: Ministry of Agriculture, Malaysia.

Ministry of Agriculture and Rural Development, Malaysia, 1975. Problems confronting the pig industry. In: T. Mahendranathan and S. Thamutaram, eds., *Proceedings of the seminar on livestock production and the food crisis held on 4th January 1975*, Kuala Lumpur: Ministry of Agriculture and Rural Development, Malaysia.

Mitloehner, F.M. and Schenker, M.B., 2007. Commentary: environmental exposure and health effects from concentrated animal feeding operations. *Epidemiology*, 18(3), pp. 309–311.

Moftah, L., 2015. Malaysia shariah law: Islamist party passes bill to implement harsh Islamic criminal punishments. *International Business Times*, [online] 19 March. Available at: <http://www.ibtimes.com/malaysia-shariah-law-islamist-party-passes-bill-implement-harsh-islamic-criminal-1852948> [Accessed 14 June 2016].

Moritz, J.S. and Latshaw, J.D., 2001. Indicators of nutritional value of hydrolyzed feathers. *Journal of Poultry Science*, 80(1), pp. 79–86.

Morris, C. and Holloway, L., 2009. Genetic technologies and the transformation of the geographies of UK livestock agriculture: a research agenda. *Progress in Human Geography*, 33(3), pp. 313–333.

Morris, C. and Kirwan, J., 2006. Vegetarians: uninvited, uncomfortable or special guests at the table of the alternative food economy? *Sociologia Ruralis*, 46(3), pp. 192–213.

Morrison, P.S., Murray, W.E. and Ngidang, D., 2006. Promoting indigenous entrepreneurship through small-scale contract farming: the poultry sector in Sarawak, Malaysia. *Singapore Journal of Tropical Geography*, 27(2), pp. 191–206.

Murray, J., 1998. Market, religion, and culture in shaker swine production, 1788–1880. *Agricultural History*, 72(3), pp. 552–573.

Myers, J., 2015. These countries eat the most meat. *World Economic Forum*, [online] 29 July. Available at: <https://www.weforum.org/agenda/2015/07/these-countries-eat-the-most-meat/> [Accessed 22 June 2016].

Nangeroni, L.I. and Kennett, P.D., 1963. *An electroencephalographic study of shechita slaughter on cortical function in ruminants.* Report prepared for the Research Institute of Religious Jewry. Ithaca: New York State Veterinary College Library Cornell University.

Nath, J., 2010. 'God is a vegetarian': the food, health and bio-spirituality of Hare Krishna, Buddhist and Seventh-Day Adventist devotees. *Health Sociology Review*, 19(3), pp. 356–368.

National Institutes of Health, 2013. Antibiotic resistance threats in the United States, 2013. US Department of Health and Human Services, Centers for Disease Control and Prevention. [online] Available at: <www.cdc.gov/drugresistance/pdf/ar-threats-2013-508.pdf> [Accessed 20 June 2016].

Nebraska Appleseed, 2009. *The speed kills you: the voice of Nebraska's meatpacking workers.* Lincoln: Nebraska Appleseed Center for Law in the Public Interest.

Nel, E. and Illgner, P., 2000. The geography of edible insects in sub-Saharan Africa: a study of the Mopane caterpillar. *The Geographical Journal*, 166(4), pp. 336–351.

Neo, H., 2009. Institutions, cultural politics and the destabilizing of the Malaysian pig industry. *Geoforum*, 40(2), pp. 260–268.

Neo, H., 2010. Geographies of subcontracting. *Geography Compass*, 4(8), pp. 1013–1024.

Neo, H., 2012. 'They hate pigs, Chinese farmers … everything!' Beastly racialization in multiethnic Malaysia. *Antipode*, 44(3), pp. 950–970.

Neo, H., 2015. Battling the head and the heart: constructing knowledgeable narratives of vegetarianism in anti-meat advocacy. In: J. Emel and H. Neo, eds., *Political ecologies of meat*. London: Routledge, pp. 236–250.

Neo, H., 2016. Ethical consumption, meaningful substitution and the challenges of vegetarianism advocacy. *Geographical Journal*, 182(2), pp. 201–212.

Neo, H. and Chen, L-H., 2009. Household income diversification and the production of local meat: the prospect of small-scale pig farming in southern Yunnan, China. *Area*, 41(3), pp. 300–309.

Neo, H. and Ngiam, J.Z., 2014. Contesting captive cetaceans: (il)legal spaces and the nature of dolphins in urban Singapore. *Social and Cultural Geography*, 15(3), pp. 235–254.

Neo, H. and Pow, C.P., 2015. Eco-cities and the promise of socio-environmental justice. In: R. Bryant, ed., *Handbook of Political Ecology*. Cheltenham: Edward Elgar, pp. 401–416.

Nestlé, M., 2013. *Food politics: how the food industry influences nutrition and health*, rev. and expanded 10th anniversary ed. Berkeley: University of California Press.

Newhook, J.C. and Blackmore, D.K., 1982. Electroencephalographic studies of stunning and slaughter of sheep and calves: the onset of permanent insensibility in calves during slaughter. *Meat Science*, 6(4), pp. 295–300.

Nguyen, T., Hermansen, J.E. and Mogensen, L., 2011. *Environmental assessment of Danish pork*. Aarhus: Det Jordbrugsvidenskabelige Fakultet, Aarhus University.

Nicholson, N.K. and Esseks, J.D., 1978. The politics of food scarcities in developing countries. *International Organization*, 32(3), pp. 679–719.

Nierenberg, D. and Reynolds, L., 2012. Disease and drought curb meat production and consumption. *Worldwatch Institute*, [online] 23 October. Available at: <http://vitalsigns.worldwatch.org/vs-trend/disease-and-drought-curb-meat-production-and-consumption> [Accessed 12 Feb 2014].

Nijdam, D., Rood, T. and Westhoek, H., 2012. The price of protein: review of land use and carbon footprints from life cycle assessments of animal food products and their substitutes. *Food Policy*, 37(6), pp. 760–770.

North, D.C., 1990. *Institutions, institutional change and economic performance*. Cambridge: Cambridge University Press.

Noske, B., 1997. *Beyond boundaries: humans and animals*. Montreal: Black Rose Books.

Novek, J., 2003a. Intensive hog farming in Manitoba: transnational treadmills and local conflicts. *Canadian Review of Sociology*, 40(1), pp. 3–26.

Novek, J., 2003b. Intensive livestock operations, disembedding, and community polarization in Manitoba. *Society and Natural Resources*, 16(7), pp. 567–581.

Ockenden, W., 2012. Footage reveals horror of Pakistani slaughter. *ABC News*, [online] 5 November. Available at: <http://www.abc.net.au/news/2012-11-05/footage-reveals-horror-of-pakistani-slaughter/4353690> [Accessed 10 April 2013].

Oh, M. and Jackson, J., 2011. Animal rights vs. cultural rights: exploring the dog meat debate in South Korea from a world polity perspective. *Journal of Intercultural Studies*, 32(1), pp. 31–56.

Oklahoma Water Resources Board [OWRB], 2007. *Oklahoma comprehensive water plan: 2007 OCWP status report*. [online] Available at: http://www.owrb.ok.gov/supply/ocwp/pdf_ocwp/WaterPlanUpdate/OCWPStatusReport2007.pdf [Accessed 25 May 2013].

Olsson, K., 2002. The shame of meatpacking. *The Nation*, [online] 29 August. Available at: <https://www.thenation.com/article/shame-meatpacking/> [Accessed 27 May 2013].

Oppel, R.A., 2013. Taping of farm cruelty is becoming the crime. *New York Times*, [online] 6 April. Available at: <http://www.nytimes.com/2013/04/07/us/taping-of-farm-cruelty-is-becoming-the-crime.html?_r=0> [Accessed 9 August 2013].

Oriental Daily News [*Dongfang Ribao*], 2015. Yulin dog meat festival triggers global internet war. [online] 12 June. Available at: <http://www.orientaldaily.com.my/international/gj5010481733> [Accessed 19 June 2016].

Pachirat, T., 2011. *Every twelve seconds: industrialized slaughter and the politics of sight*. New Haven: Yale University Press.

Pence, G.E., ed. 2002. *The ethics of food: a reader for the twenty-first century*. Lanham, MD: Rowman and Littlefield.

Petersen, M., 2003. Indians now disdain a farm once hailed for giving tribe jobs. *New York Times*, [online] 15 November. Available at: <http://www.nytimes.com/2003/11/15/us/indians-now-disdain-a-farm-once-hailed-for-giving-tribe-jobs.html> [Accessed 7 August 2013].

Petersen, M., 2012. As beef cattle become behemoths, who are animal scientists serving? *Chronicle of Higher Education*, [online] 15 April. Available at: <http://chronicle.com/a rticle/As-Beef-Cattle-Become/131480/> [Accessed 27 May 2014].

Pew Commission on Industrial Farm Animal Production [Pew Commission], 2009. *Putting meat on the table: industrial farm animal production in America.* A report of the Pew Commission on Industrial Farm Animal Production. [online] Available at: <http://www.ncifap.org/_images/PCIFAPFin.pdf> [Accessed 20 June 2016].

Phelps, N.A. and Wood, A., 2006. Lost in translation? Local interests, global actors and the multi-scalar dynamics of inward investment. *Journal of Economic Geography*, 6(4), pp. 493–515.

Philipkoski, K., 2003. Sour grapes over milk labeling. *Wired*, [online], 16 September. Available at: <http://archive.wired.com/medtech/health/news/2003/09/60132?curren tPage=all> [Accessed 1 July 2013].

Philo, C. and Wilbert, C., 2000. Animal spaces and beastly places: an introduction. In: C. Philo and C. Wilbert, eds., *Animal spaces, beastly places: new geographies of animal–human relations.* London: Routledge.

Piazza, J., Ruby, M.B., Loughnan, S., Luong, M., Kulik, J., Watkins, H.M. and Seigerman, M., 2015. Rationalizing meat consumption: the 4Ns, *Appetite*, 91(1), pp. 114–128.

Pig International, 2005. Profiling China. *Pig International*, 35(1), pp. 12–13.

Pig International, 2007. China's biggest for pork. *Pig International*, 37(1), pp. 12–19.

Pingali, P. and McCullough, E., 2010. Drivers of change in global agriculture and livestock systems. In: H. Steinfeld, H.A. Mooney, F. Schneider and L.E. Neville, eds., *Livestock in a changing landscape*, vol. 1: *Drivers, consequences, and responses.* Washington: Island Press, pp. 5–10.

Pisano, G.P. and Shih, W.C., 2009. Restoring American competitiveness. Available at: <https://hbr.org/2009/07/restoring-american-competitiveness> [Accessed 7 November 2016].

Pluhar, E.B., 2010. Meat and morality: alternatives to factory farming. *Journal of Agriculture and Environment Ethics*, 23(5), pp. 455–468.

Popp, A., Lotze-Campen, H. and Bodirsky, B., 2010. Food consumption, diet shifts and associated non-CO_2 greenhouse gases from agricultural production. *Global Environmental Change*, 20(3), pp. 451–462.

Pray, C., Gisselquist, D. and Nagarajan, L., 2011. Private investment in agricultural research and technology transfer in Africa. Conference Working Paper 13. In: *IFPRI-ASTI/FARA Conference*. Accra, Ghana, 5–7 December, [online] Available at: <http://www.asti.cgiar.org/pdf/conference/Theme4/Pray.pdf> [Accessed 27 May 2014].

Pritchard, B., 2000. Geographies of the firm and transnational agro-food corporations in East Asia. *Singapore Journal of Tropical Geography*, 21(3), pp. 246–262.

Pritchard, W.N., 1998. The emerging contours of the third food regime: evidence from Australian dairy and wheat sectors. *Economic Geography*, 74(1), pp. 64–74.

Prola, L., Nery, J., Lauwaerts, A., Bianchi, C., Sterpone, L., De Marco, M., Pozzo, L. and Schiavone, A., 2013. Effects of N,N-dimethylglycine sodium salt on apparent digestibility, vitamin E absorption, and serum proteins in broiler chickens fed a high- or low-fat diet. *Poultry Science*, 92(5), pp. 1221–1226.

Raloff, J., 2002. Hormones: here's the beef: environmental concerns reemerge over steroids given to livestock. *Science News*, 161(1), pp. 10–12.

Randalls, S., 2011. Broadening debates on climate change ethics: beyond carbon calculation. *The Geographical Journal*, 177(2), pp. 127–137.

Raymond, R., Bales, C.W., Bauman, D.E., Clemmons, D., Kleinman, R., Lanna, D., Nickerson, S. and Sejrsen, K., 2009. Recombinant bovine somatotropin (rBST): a safety assessment. In: *Joint Annual Meeting, American Dairy Science Association, Canadian Society of Anim Sci, and American Society of Anim Sci*, Montreal, Canada, [online] 14 July. Available at: <http://www.naiaonline.org/pdfs/Recombina ntSomatotropinASafetyAssessment2010.pdf> [Accessed 23 May 2014].

Reid, J., 2014. Climate, migration, and sex: the biopolitics of climate-induced migration. *Critical Studies on Security*, 2(2), pp. 196–209.

Ritzer, G. and Jurgenson, N., 2010. Production, consumption, prosumption: the nature of capitalism in the age of the digital 'prosumer'. *Journal of Consumer Culture*, 10(1), pp. 13–36.

Robbins, P., 1998. Shrines and butchers: animals as deities, capital, and meat in contemporary north India. In: J. Wolch and J. Emel, eds., *Animal geographies: place, politics, and identity in the nature-culture borderlands*. London: Verso, pp. 218–240.

Roe, E.J., 2006. Things becoming food and the embodied, material practices of an organic food consumer. *Sociologia Ruralis*, 46(2), pp. 104–121.

Rose, N., 2006. Governing 'advanced' liberal democracies. In: A. Sharma and A. Gupta, eds., *The anthropology of the state: a reader*. Oxford: Blackwell, pp. 142–162.

Rose, N., 2013. The human sciences in a biological age. *Theory, Culture and Society*, 30(1), pp. 3–34.

Rosin, C. and Cooper, M.H., 2015. Mitigating greenhouse gas emissions from livestock: complications, implication and new political ecology. In: J. Emel and H. Neo, eds., *Political ecologies of meat*. London: Routledge, pp. 315–328.

Rouse, C. and Hoskins, J., 2004. Purity, soul food, and Sunni Islam: explorations at the intersection of consumption and resistance. *Cultural Anthropology*, 19(2), pp. 226–249.

Rubenstein, M., 1931. Relations of science, technology, and economics under capitalism, and in the Soviet Union. In: *Science at the Crossroads: Papers Presented to the International Congress of the History of Science and Technology*, London, [online] Available at: <https://www.marxists.org/subject/science/essays/rubinstein.htm> [Accessed 23 May 2014].

Salmi, P., 2005. Rural pluriactivity as a coping strategy in small-scale fisheries. *Sociologia Ruralis*, 45(1/2), pp. 22–36.

Scales-Trent, J., 2001. Racial purity laws in the United States and Nazi Germany: the targeting process. *Humans Rights Quarterly*, 23(2), pp. 259–307.

Schlosser, E., 2001. *Fast food nation: the dark side of the all-American meal*. Boston: Houghton Mifflin.

Schmalzer, S., 2002. Breeding a better China: pigs, practices, and place in a Chinese county, 1929–1937. *Geographical Review*, 92(1), pp. 1–22.

Schmidt, C., 2009. Swine CAFOs & novel H1N1 flu: separating facts from fears. *Environmental Health Perspectives*, 117(9), pp. A394–A401.

Schneider, K., 1990. Betting the farm on biotech. *New York Times*, [online] 10 June. Available at: <http://www.nytimes.com/1990/06/10/magazine/betting-the-farm -on-biotech.html> [Accessed 23 May 2014].

Schneider, M., 2011. *Feeding China's pigs: implications for the environment, China's smallholder farmers and food security*. Minneapolis: Institute for Agriculture and Trade Policy, [online] Available at: <http://www.iatp.org/files/2011_04_25_Feeding ChinasPigs_0.pdf> [Accessed 27 May 2014].

Schneider, M. and Sharma, S., 2014. *China's pork miracle? Agribusiness and development in China's pork industry.* Minneapolis: Institute for Agriculture and Trade Policy, [online] Available at: <www.iatp.org/files/2014_03_26_PorkReport_f_web. pdf> [Accessed 27 May 2014].

Schoffeleers, J.M. and Meijers, D., 1978. *Religion, nationalism and economic action: critical questions on Durkheim and Weber.* Assen: Van Gorcum.

Schurman, R.A., 1996. Snails, southern hake and sustainability: neoliberalism and natural resource exports in Chile. *World Development*, 24(11), pp. 1695–1709.

Scott, J.C., 1976. *The moral economy of the peasant: rebellion and subsistence in Southeast Asia.* New Haven: Yale University Press.

Shaw, B.E., 1940. Geography of mast feeding. *Economic Geography*, 16(3), pp. 239–249.

Shimshony, A. and Chaudry, M.M., 2005. Slaughter of animals for human consumption. *Review of Science and Technology*, 24(2), pp. 693–710.

Shukin, N., 2009. *Animal capital: rendering life in biopolitical times.* Minneapolis: University of Minnesota Press.

Smart, A., 2004. Adrift in the mainstream: challenges facing the UK vegetarian movement. *British Food Journal*, 106(2), pp. 79–92.

Smith, A., Stirling, A. and Berkhout, F., 2005. The governance of sustainable socio-technical transitions. *Research Policy*, 34(10), pp. 1491–1510.

Smith, T.C., Male, M.J., Harper, A.L., Kroeger, J.S., Tinkler, G.P., Moritz, E.D., Capuano, A.W., Herwaldt, L.A. and Diekema, D.J., 2009. Methicillin-resistant *Staphylococcus aureus* (MRSA) strain ST 398 is present in midwestern U.S. swine and swine workers. *PLoS ONE*, 4(1).

Smithfield Foods, 2014. *Sustainability and Financial Report 2014.* [online] Available at: <http://www.smithfieldfoods.com/integrated-report/introduction> [Accessed 14 June 2015].

Sneeringer, S., 2009. Does animal feeding operation pollution hurt public health? A national longitudinal study of health externalities identified by geographic shifts in livestock production. *American Journal of Agricultural Economics*, 91(1), pp. 124–137.

Sneeringer, S., MacDonald, J.M., Key, N., McBride, W.D. and Mathews, K., 2015. *Economics of antibiotic use in U.S. livestock production.* Economic Research Report no. 200, United States Department of Agriculture.

St Martin, K., 2006. The impact of 'community' on fisheries management in the US northeast. *Geoforum*, 37(2), pp. 169–184.

Steinfeld, H., Gerber, P., Wassenaar, T., Castel, V., Rosales, M. and de Haan, C., 2006. *Livestock's long shadow: environmental issues and options.* Rome: Food and Agriculture Organization/LEAD.

Storper, M., 1989. The transition to flexible specialisation in the US film industry: external economies, the division of labour, and the crossing of industrial divides. *Cambridge Journal of Economics*, 13(2), pp. 273–305.

Storper, M., 2009. Roepke lecture in economic geography – regional context and global trade. *Economic Geography*, 85(1), pp. 1–21.

Stringer, C., Simmons, G. and Rees, E., 2011. Shifting post-production patterns: exploring changes in New Zealand's seafood processing industry. *New Zealand Geographer*, 67(3), pp. 161–173.

Stull, D. and Broadway, M.J., 2004. *Slaughterhouse blues: the meat and poultry industry in America*, Belmont, CA: Wadsworth Learning.

Subak, S., 1999. Global environmental costs of beef production. *Ecological Economics*, 30(1), pp. 79–91.

Suresh, A., Choi, H.L., Oh, D.I. and Moon, O.K., 2009. Prediction of the nutrients value and biochemical characteristics of swine slurry by measurement of EC-electrical conductivity. *Bioresource Technology*, 100(20), pp. 4683–4689.

Talbot, J.M., 1995. The regulation of the world coffee market: tropical commodities and the limits of globalization. In: P. McMichael, ed., *Food and agrarian orders in the world-economy*, Westport, CT: Greenwood Press, pp. 139–168.

Tam, L., 2011. Measuring San Francisco's ecological footprint. *SPUR*, [online] 5 July. Available at: <http://www.spur.org/news/2011-07-05/measuring-san-franciscos-ecolo gical-footprint> [Accessed 16 June 2016].

Tan, S.B.H., 2000. Coffee frontiers in the Central Highlands of Vietnam: networks of connectivity. *Asia Pacific Viewpoint*, 41(1), pp. 51–67.

Tanaka, K. and Juska, A., 2010. Technoscience in agriculture: reflections on the contributions of the MSU School of Sociology of Food and Agriculture. *Journal of Rural Social Sciences*, 25(3), pp. 34–55.

Taylor, A.L., 2012. Rise of the 'semi-vegetarians'. <http://www.bbc.co.uk/food/0/ 19294585> [Accessed 7 November 2016].

Terzić, A. and Bjeljac, Z., 2014. The extermination of Jewish population and heritage in Bačka region of Serbia. *Trames: Journal of the Humanities and Social Sciences*, 18(1), pp. 39–55.

Thorne, P.S., 2002. Air quality issues. In: *Iowa concentrated animal feeding operations air quality study*. Iowa City, IA: ISU/UI Study Group, University of Iowa College of Public Health, pp. 35–44.

Thornton, P.K., 2010. Livestock production: Recent trends, future prospects. *Philosophical Transactions of the Royal Society of London. Series B, Biological Sciences*, 365(1554), pp. 2853–2867.

Tobler, C., Visschers, V.H.M. and Siegrist, M., 2011. Eating green: consumers' willingness to adopt ecological food consumption. *Appetite*, 57(3), pp. 674–682.

Torres, B., 2007. *Making a killing: the political economy of animal rights*, Oakland, CA: AK Press.

Tserendejid, Z., Hwang, J., Lee, J. and Park, H., 2013. The consumption of more vegetables and less meat is associated with higher levels of acculturation among Mongolians in South Korea. *Nutrition Research*, 33(12), pp. 1019–1025.

Twine, R., 2010. *Animals as biotechnology: ethics, sustainability and critical animal studies*. London: Earthscan.

Tyson Foods, 2016. *Facts about Tyson Foods*. [online] Available at: <http://ir.tyson. com/investor-relations/investor-overview/tyson-factbook/default.aspx> [Accessed 20 May 2016].

Ufkes, F.M., 1998. Building a better pig: Fat profits in lean meat. In: J. Wolch and J. Emel, eds., *Animal geographies: place, politics, and identity in the nature-culture borderlands*. London: Verso, pp. 241–255.

United Nations Development Programme [UNDP], 1976. *Report in the livestock development survey – Malaysia (1976)*. Rome: United Nations Development Programme and Food and Agriculture Organisation.

US Department of Agriculture [USDA], 2010. *Overview of the United States dairy industry*. National Agricultural Statistics Service (NASS), Agricultural Statistics Board. [online] Available at: <http://usda.mannlib.cornell.edu/usda/current/USDa iryIndus/USDairyIndus-09-22-2010.pdf> [Accessed 23 May 2014].

US Department of Agriculture [USDA], 2014. *2012 census of agriculture*. US Department of Agriculture. [online] Available at: <https://agcensus.usda.gov/Publications/2012/#full_report> [Accessed 15 May 2015].

US Department of Agriculture [USDA], 2015. *2015–2020 Dietary Guidelines for Americans* 8th ed. Report by the US Department of Agriculture, Center for Nutrition Policy and Promotion. [online] Available at: <http://www.cnpp.usda.gov/dietary-guidelines> [Accessed 19 June 2016].

US Government Accountability Office [GAO], 2010. *Humane Methods of Slaughter Act: actions are needed to strengthen enforcement*. Report to Congressional Requesters, GAO-10-203, [online] February. Available at: <http://www.gao.gov/products/GAO-10-203> [Accessed 19 June 2016].

Vasek, L., 2011. Ban on live cattle trade to Indonesia to last up to six months, Joe Ludwig announces. *The Australian*, [online] 8 June. Available at: <http://www.theaustralian.com.au/national-affairs/temporary-ban-on-exports-leaves-cattle-stranded/story-fn59niix-1226071488817> [Accessed 10 April 2013].

Veeck, G., 2013. China's food security: past success and future challenges. *Eurasian Geography and Economics*, 54(1), pp. 42–56.

Vegetarian Butcher, The (n.d.). About us. [online] Available at: <http://www.vegetarianbutcher.com/about-us/the-vegetarian-butcher> [Accessed 20 May 2016].

Vos, B.I., de and Bush, S.R., 2011. Far more than market-based: rethinking the impact of the Dutch Viswijzer (Good Fish Guide) on fisheries' governance. *Sociologia Ruralis*, 51(3), pp. 284–303.

Vries, M., de and de Boer, I.J.M., 2010. Comparing environmental impacts for livestock products: a review of life cycle assessments. *Livestock Science*, 128(1/3), pp. 1–11.

Wackernagel, M. and Rees, W.E., 1996. *Our ecological footprint: reducing human impacts on the Earth*. Gabriola Island, BC: New Society Publishers.

Waithanji, E., 2015. The political ecology of factory farming in East Africa. In: J. Emel and H. Neo, eds., *Political ecologies of meat*. London: Routledge, pp. 67–83.

Walkenhorst, P., 2004. Economic transition and the sectoral patterns of foreign direct investment. *Emerging Markets Finance and Trade*, 40(2), pp. 5–26.

Walker, P., Rhubart-Berg, P., McKenzie, S., Kelling, K. and Lawrence, R.S., 2005. Public health implications of meat production and consumption. *Public Health Nutrition*, 8(4), pp. 348–356.

Wallerstein, I., 1974. The rise and future demise of the world capitalist system: concepts for comparative analysis. *Comparative Studies in Society and History*, 16(4), pp. 387–415.

Warrick, T., 2001. 'They die piece by piece': in overtaxed plants, humane treatment of cattle is often a battle lost. *Washington Post*, 10 April, p. 1.

Watts, M.J., 1994. Life under contract: contract farming, agrarian restructuring, and flexible accumulation. In: P.D. Little, and M.J. Watts, eds., *Living under contract: contract farming and agrarian transformation in sub-Saharan Africa*, Madison: University of Wisconsin Press, pp. 21–77.

Weber, C.L. and Matthews, H.S., 2008. Food-miles and the relative climate impacts of food choices in the United States. *Environmental Science Technology*, 42(10), pp. 3508–3513.

Weis, T., 2013. The meat of the global food crisis. *Journal of Peasant Studies*, 40(1), pp. 65–85.

Weiss, T.G. and Jordan, R.S., 1976. *The world food conference and global problem solving*. New York: Praeger.

Whatmore, S., 1994. Global agro-food complexes and the refashioning of rural Europe. In: A. Amin and N. Thrift, eds., *Globalization, institutions, and regional development in Europe*. Oxford: Oxford University Press, pp. 46–67.

Whatmore, S., 2002. From farming to agribusiness: the global agro-food system. In: R.J. Johnston, P.T. Taylor and M.J. Watts, eds., *Geographies of global change: remapping the world*, Oxford: Blackwell, pp. 57–67.

Whitaker, A., 1980. Pesticide use in early twentieth century animal disease control. *Agricultural History*, 54(1), pp. 71–81.

Wilbert, C., 2007. The birds, the birds: biopolitics and biosecurity in the contested spaces of avian flu. In L. Davis, ed., *Forum on Contemporary Art & Society (Focas) 6: Regional animalities: cultures, natures, humans & animals in Southeast Asia*. Singapore: NUS Press, pp.102–123.

Wilbert, C., 2009. Animal geographies. In: R. Kitchin and T. Nigel, eds., *International encyclopedia of human geography*. Oxford: Elsevier, pp. 122–126.

Willerton, A. and Proulx, M., 2012. Glue of the future. 25 October. [online] Available at: <http://www.ales.ualberta.ca/ALESNews/2012/October/Glueofthefuture.aspx> [Accessed 28 May 2014].

Williams, A., 2004. Disciplining animals: sentience, production, critique. *International Journal of Sociology and Social Policy*, 24(9), pp. 45–57.

Wilson, J., 1986. The political economy of contract farming. *Review of Radical Political Economics*, 18(4), pp. 47–70.

Wilson, J., 2000. Rosebud hog wrangle. *South Dakota Magazine*, September–October, pp. 8–16.

Wolch, J., Griffith, M. and Lassiter, U., 2002. Animal practices and the racialization of Filipinas in Los Angeles. *Society and Animals*, 10(3), pp. 221–248.

Wolch, J. and Emel, J., eds., 1998. *Animal geographies: place, politics, and identity in the nature-culture borderlands*. London: Verso.

Wolfe, C., 2013. *Before the law: humans and other animals in a biopolitical frame*. Chicago: University of Chicago Press.

World Bank, 2009. *Minding the stock: bringing public policy to bear on livestock sector development*. Report no. 44010-GLB, Washington, DC: World Bank.

World Conference on Animal Production [WCAP], 2013. The 11th World Conference on Animal Production, Beijing, 15–20 October 2013.

World Organisation for Animal Health [OIE], 2014. *New OIE international standards and guidelines on animal health*. [online] 25–30 May. Available at: <http://www.oie.int/for-the-media/press-releases/detail/article/new-oie-international-standards-and-guidelines-on-animal-health/> [Accessed 1 June 2014].

Worldwatch Institute, 2013. Global meat production and consumption continue to rise. [online] Available at: <http://www.worldwatch.org/global-meat-production-and-consumption-continue-rise-1> [Accessed 10 October 2013].

Woś, A., 1994. Państwowa gospodarka w rolnictwie w okresie transformacji systemowej. *Wieś i Rolnictwo*, 1994/2, pp. 26–40.

Wrigley, N., 2002. Transforming the corporate landscape of US food retailing: market power, financial re-engineering and regulation. *Tijdschrift voor Economische en Sociale Geografie*, 93(1), pp. 62–82.

Wu, F.H., 2002. The best 'chink' food: dog eating and the dilemma of diversity. *Gastronomica*, 2(2), pp. 38–45.

Yarwood, R. and Evans, N., 2006. A Lleyn sweep for local sheep? Breed societies and the geographies of Welsh livestock. *Environment and Planning A*, 38(7), pp. 1307–1326.

Ye, L.-S., 2003. *The Chinese dilemma*. Kingsford, NSW: East West Publishing.

Yin, J., 2006. China's pig profile in numbers. *Pig International*, September, pp. 22–23.

Young, C. and Kaczmarek, S., 2000. Local government, local economic development and quality of life in Poland. *GeoJournal*, 50(2/3), pp. 225–234.

Young, T., 2006. Recycled chickens: farmers turn to composting amidst collapsed spent-hen market. *The Press Democrat*. [online] 22 November. Available at: http://www.pressdemocrat.com/article/20061122/NEWS/611220399 [Accessed 7 October 2013].

Yu, H., Jin, F., Shu, X.O., Li, B.D., Dai, Q., Cheng, J.R., Berkel, H.J. and Zheng, W., 2002. Insulin-like growth factors and breast cancer risk in Chinese women. *Cancer Epidemiology, Biomarkers, and Prevention*, 11(8), pp. 705–712.

Zaid, 2000. Penternak babi harus sedar risiko wabak JE. *Utusan Malaysia*, [online] 27 June. Available at: <http://ww1.utusan.com.my/utusan/info.asp?y=2000&dt=0627&pub=Utusan_Malaysia&sec=Forum&pg=fo_04.htm> [Accessed 12 March 2012].

Zheng, P., 1985. *Livestock breeds of China*. Animal Production and Health Papers, vol. 46. Rome: Food and Agriculture Organisation of the United Nations.

Index

abattoirs *see* slaughterhouses
ABS Global 43
activists: animal rights 130; animal
    welfare 2; climate 2; vegetarianism
    117, 119, 123, 126, 133
Africa 42, 83, 117
ag-gag laws 72, 95
Agri Plus 25–7, *29*, 29–32
agricultural 49; colleges 42; commodities
    46; development 15; extension services
    51; geography 13; land 24, 39; markets
    14; policies 14; production 21; science
    42, 47, 52
Agricultural Department (Jinghong)
    110–11, 113–16
agrifood 10, 13–15, 17; conglomerate 18;
    industry 16; systems 46
Agrium 66
air pollution 68–9, 76, 84, 95
Alberta Agriculture and Rural
    Development Ministry 57
Amazon 64, 68
American Cyanamid 47
American Dairy Science Association 50
American Society for the Prevention of
    Cruelty to Animals 90
American Society of Animal Nutrition 50
American Veterinary Medical
    Association (AVMA) 56, 85
American(s) 25–6, 80, 133; catfish
    industry 18; livestock industry 41, 46,
    50; pork consumption 109; *see also*
    North American
Americas 4, 6
Animal Aid 97
Animal Enterprise Terrorism Act 72
Animal machines (Harrison) 87–8
animal(s): biotechnologies 46; cruelty to
    62, 72, 76, 95, 97, 99–100, 102,
118–19; feeding 1, 4, 27, 54, 60, 72,
    78, 80, 84–5, 131; food 1–6, 8–12, 14,
    18, 23, 40, 59–65, 68, 82, 84, 86–8,
    98–9, 101, 106–9, 117, 119, 124,
    127–9, 132–5, 137–40, 142; genetics
    42, 51–2, 109; geographies of 1–3, *3*, 5,
    8–10, 12; health 46, 52, 88, 103, 138;
    -industrial complex 86–7; waste 42
animal rights 98, 121, 124, 126, 130;
    activists 130; group 128; movement
    119; organisations 72
animal science 42, 46, 48, 50–2, 61,
    89–90, 129, 142; industry 4; research
    41; technology 6
animal welfare 4, 41–2, 52, 57, 65, 67,
    74–5, 77–8, 83–4, 87–92, 98–9, 101–3,
    106, 117–21, 124, **125**, 127, 139;
    activists 2; groups 56, 98, 141; in
    Indonesia 99, 101–2; policies 75, 88;
    regulation 107
Animal Welfare Act 96
Animals and Society Institute 89–90
Animals Australia 99, 101
Animex 26, 31
anti-meat activism 131, 133
aquaculture 77–8, 138–9
Asia 4, 6, 43, 141; aquaculture in 138;
    meat production in 107
Asian Development Bank 139
Attorney General's Office (USA) 95
Atwood, Margaret 129
Australia 4, 48, 72, 99, 102; and animal
    welfare 101; livestock exports from 66,
    101–3; slaughterhouses in 76, 105
Australian 99–100, 102; government
    99, 102–3; meat consumption 109;
    states 85
Australian Livestock Export
    Corporation 99

authoritarianism 33
Avantea 129
Aviagen Broilers 43

Babjee, Ahmad Mustaffa 39
Bahrain 102–3
Barrera, Anthony 94
beef 22, 44, 54, 69, 71, 81, 132, 140–1;
  consumption 65, 81; exports 66;
  industry 102; packing 65–6; producers
  63; production 64; research 43, 51;
  slaughterhouse 75
Belgium 142; as pig producer 24
Bell Farms Group 91–2, 95
Beltsville 50
Beyond Beef 132
Beyond Chicken 132–3
Beyond Meat 132–3
biodiversity 68, 70–1; in China 109;
  loss 76, 78
bioengineering 129
biofuel companies 57
biopolitics 1, *3*, 4–5, 7–8, 12, 41, 44–5,
  64; of consumption 8; of food animals
  3, 142; of population 86; *see also*
  politics
biopower 7, 44
Bíos: biopolitics and philosophy
  (Esposito) 86
Bom, Paul 132
Botswana 70
bovine: growth hormone 47, 70;
  tuberculosis 137–8
bovine spongiform encephalopathy
  (BSE) 57–8, 137
Brambell Committee 88–9; report 88
Brazil 6, 54, 66, 117; livestock emissions
  79; pig industry 83; R&D companies
  in 52
Brazilian: meatpacking company 44, 66;
  research 55
breeding 45–6, 50–2, 54, 59–60; cattle
  43, 54; dairy 51, 54; livestock 50;
  pig 27, 31, 36, 43, 55, 109, 128;
  poultry 43
Bresky, Steven J. 72
Breuil 56
BRF 6
Brin, Sergey 1
Britain 31, 57, 66, 74, 97, 142; analogue
  meat in 132; animal welfare in 88;
  BSE in 58; foot-and-mouth disease in
  58; food policy in 18
British government 88, 97

British and Foreign Society for the
  Promotion of Humanity and
  Abstinence from Animal Food 116
Brown, Alton 133
Buddhists 140
Buffalo Lake National Wildlife Refuge 71
Bulgaria 79
bumiputera 35–6
Bureau of Animal Industry (USA) 50
Bureau of Labor Statistics (USA) 75
Burnt Thighs Nation *see* Rosebud Sioux
Bushway Packing 97

Cactus Feeders Inc. 22
Cambridge Declaration on Conscious-
  ness 90
Canada 48, 73–4; concentrated (con-
  fined) animal feed operation(s) in 84
Canadian Veterinary Medical Associa-
  tion 85
capitalism 17, 49, 87; state 45
Cargill 6
Carson, Rachel 87
cattle 2, 7, 51, 53, 59, 64, 66, 70–1, 75,
  84–5, 97–9; Australian 99, 102; beef
  54; breeding 43, 50, 54, 59–60; cloned
  128; dairy 43, 54; Indonesia 99–100;
  Italian 108; production 69
Centers for Disease Control and
  Prevention (USA) 80
Central America 71
Central Bank of China 17
Central Europe 24, 44, 66
Central Valley Meat Company 96
CF Industries 66
Chatham House 79
chickens 6, 53, 55, 57, 62, 90, 97, 129; in
  USA 75, 83, 85, 96
China 44–5, 52–3, 66, 74, 108–10, *111*,
  139, 141; contract farming in 21; food
  safety in 66; livestock production in
  70, 110; meat consumption in 109,
  115; meat production in 135; pig
  farming in 43, 66, 83, 108–9, *112–13*,
  115; pork in 5, 18, 66, 109
China Agricultural University College of
  Animal Science and Technology 51
Chinese 110, 141; companies 66;
  government 5, 17–18, 143; in Malaysia
  33, 35–9; pig farming 33, 35–8; pig
  industry 66, 110; pork policy 18; pork
  processor 18
climate 77; activists 2; change 76, 131;
  -induced migration 7

cloning 78, 128
Cobb-Vantress 43
coffee 15
colonialism 15, 51
commodification: of food animals 1, *3*,
    3–5, 7–8, 10–12, 18, 21, 23, 41, 52, 59,
    61, 63–5, 80, 115, 135, 142; of meat
    140; seed 46
companies 21–3, 26, 30–2, 48, 50, 52,
    95–6, 118, 148; animal chemical 48;
    biofuel 57; aquaculture 138; breeding
    51; Chinese 66; cloning 128–9; Danish
    25–6; Dutch 132; European 44; feed
    42, 49; genetics 42–3, 49; global 31,
    40, 65; livestock export 103; meat 19,
    22, 24, 26, 44, 65–7, 136; meat
    analogue 143; meatpacking 6, 21, 44,
    65–6, 131, 137; multinational 15, 52;
    North American 44; pharmaceutical
    42; pig 66; poultry 75; seed 46;
    transnational 14
concentrated (confined) animal feed
    operation(s) 58–9, 83–6, 88–9, *92*, 98,
    106, 118–19; and pollution 69, 71;
    workers 71–4, 137; *see also* feed
conglomerate 6; agrofood 4, 65; meat
    44, 65
Conover, Ted 95
consciousness: of animals 90, 103, 105;
    Islamic 38
consumer(s) 2, 7–8, 13, 18, 37, 41, 45,
    47–8, 59, 78, 82, 86, 127, 138, 142;
    demand 81; and food animals 5,
    10–11, 46, 61, 84, 106, 119, 127,
    131, 134–6, 139, 142; and food taboos
    141; meat 6, 10, 67, 115–18, 120–3,
    127–8, 131, 133, 136; moral 127–8;
    movements 2; of pork 66, 109, 114;
    rights 78
Consumers' Association of Penang 81
contract farmers 6, 21, 67, 136; in Poland
    29, 31
contract farming 18, 20–3, 44, 65, 119; in
    China 21; in East Malaysia 22–3; in
    Poland 24, 27, 30–1; in USA 20, 22
corporate social responsibility 18, 31–2;
    and meat industry 20, 23; Poldanor
    and 32
Cove, The 141
critical animal geographies 1–3, 5, 9–10, 12
cropland 68
cultural politics 33, 40, 81, 139; of
    consumption 14; and meat 78; of pig
    production 135

culture 2–3, 7–9, 11, 20, 32–3, 45, 116,
    140, 142; of food consumption 15,
    109, 115–16, 133, 141; of food
    production 8, 115; in Malaysia 32,
    36, 40; rural 67; and vegetarianism
    116–7, 127
cultured meat 128–9
Czech Republic 24

dairy industry 48, 51, 78, 136
DanBred 43
Danish: company 25–6; investors 27; pig
    producers 54–5, 71
Danish Crown AmbA 6,44
deforestation 64, 69–70, 76, 79
Delta Darlene 128
democracy 46
Denmark 31, 43, 74, 85, 128; as pig
    producer 24
Department of Agriculture (USA) 43, 46,
    48, 50, 52, 74–5, 96, 98, 118, 128
Department of Veterinary Services
    (Malaysia) 34
desertification 76
development: rural 13, 36, 67,
    108, 138
developmental: organisations 4; strategy
    32, 110
Discipline and punishment (Foucault) 59

East Asia 120; meat consumption in 116
Eastern Europe 44, 66, 83; investors in
    24, 66; workers from 71
Eating animals (Foer) 83
ecological: degradation 46; footprint 77;
    hoofprint 76; impacts of livestock
    production 64, 67–9; systems 76
economic development 5, 20, 22, 24; in
    livestock industry 65
economy 3, 7, 13, 61, 80, 142; Australian
    102; capitalist 14, 46; food 6, 108;
    household 139; local 19, 22, 27;
    market 5–6, 25, 44, 66; *see also*
    political economy
Elanco 48
Eli Lilly 47
employment 27; female 15; loss of 74;
    rural 25, 27
energy depletion 76
England 19, 85
environment 4, 17, 63, 65, 76, 82, 110,
    120, 123–4, 127, 133–4, 142–3;
    damage to 68; impacts on 1, 68–9,
    122, 138; protecting 121

environmental 23, 42, 121, 134; benefits
126; concerns 40m, 108; degradation
119; impacts 4, 12, 64–5, 68, 76–8, 80,
82, 106, 119, 121–2, 137; movement
117, 119–20; performance 34;
pollution 12, 34; problems 67;
protection 121; sustainability 55;
studies 9; technologies 107; welfare 41
environmentalism 119
Enviropig 79
Esposito, Roberto 83, 86–7
ethology 89, 91
Europe 4, 6, 14, 41, 51, 74; agricultural
policies in 14; meat scandal in 141;
see also Central Europe *and* Eastern
Europe
European Parliament 91
European Union 2, 24, 48, 52, 54, 58,
98, 128
Every twelve seconds (Pachirat) 83
EW Group 43
Exporter Supply Chain Assurance
System 103

factory farm(s) 1, 4, 88; workers in 67, 71
factory farming 1, 3, 67, 107–8, 118–19,
125; environmental impact of 76, 80;
pig 40, 71, 74, 113
famine 5
Farm Animal Welfare Council (UK)
88–9
farm(s) 6, 25, *29*, 42, 65, 78, 106; British
58; Chinese 17, 33–5, 66, 115; contract
22–3, 27, 65; dairy 46–7; factory 1, 4,
67, 71, 88; industrial 39; livestock
67, 136; Malaysian 33–5, 37–9, 81; pig
6, 24–5, 27, 29, 32–5, 37–9, 64, 66, 71,
81; Polish 27, 29, 31–2, 39; small 22,
24; state-owned 25–6, 32; in USA
46–8, 64, 71–3
farmer(s) 8, 14, 20, 25, 27, 29–30, 32,
41, 43–4, 46–8, 50, 60, 63–4, 73,
80–2, 138; Australian 102; cattle 99;
Chinese 33–9, 114; and companies 21;
contract 6, 21–3, *29*, 30–2, 67, 136;
cooperatives 43; and cultural norms
23; dairy 48, 62; in developing
countries 91; fish 138–9; and local
government 21; pig 34–9, 62, 73,
81, 114; small-scale 12, 21, 49, 60;
tenant 62
farming: contract 18, 20–4, 27, 30–1, 44,
65, 119; factory 1, 3, 40, 67, 71, 74,
76, 80, 107–8, 113, 118–19, **125**;

intensified 1–5, 61; livestock 1, 3–4,
13, 106, 108, 110, 115, 119, 134, 138;
organic 18, 107–10, 133; pig 4, 17, 20,
24, 33–9, 43, 108; poultry 62
Fast food nation (Schlosser) 83
Fats and Proteins Research Foundation 58
Federation of Animal Science Societies 56
Federation of Livestock Farmers of
Malaysia 81
feed 21, *28*, 30, 42, 49, 52–3, 57–9, 65,
69, 73, 78, 80, 85, 113, *114*, 129;
companies 42; control 45; crops 68;
production 70; *see also* concentrated
(confined) animal feed operation(s)
Fine, Ben 16–17, 19
fish farmer(s) 138–9
fishery industry 139
food: animals 1–6, 8–12, 14, 18, 23, 40,
59–65, 68, 82, 84, 86–8, 98–9, 101,
106–9, 117, 119, 124, 127–9, 132–5,
137–40, 142; geographies of 7, 13;
justice 3–4, 9; political economy of 13,
16–17; quality 2, 7, 14, 26; regime
13–17; safety 14, 37, 66–7, 74–5, 80,
82, 97–8, 108, 131, 133, 139; security
3, 5–6, 18, 139; taboos 140–1;
underproduction of 5
Food and Drug Administration (United
States) 47–8, 97, 128
Food Safety and Inspection Service
(USA) 97
foot-and-mouth disease 51, 58
Ford, Henry 55
Fordist regime 67, 135
foreign direct investment 24; in East
Central Europe 24; in Poland 14,
18–19, 24; transnational 19–21, 44
Foucault, Michel 7, 41–2, 44–5, 52,
59, 86
Four Corners 99
France 49, 73–4, 79, 98, 142
free-range: meat 77; methods 77; pigs 86
Friends of the Earth Europe 135

Garrett, Jennifer 48
gender 45, 140
Genentech 47
genetics 44, 51, 53–4; animal 42,
51–2, 109; companies 43, 49;
livestock 51, 55; pig 43, 71, 79;
poultry 49, 52
genomic selection 54
genomics 41
Genus 43

geography 124, 134–5; agricultural 13; animals 1–3, 5, 8–10, 12; economic 138; of food 7, 13; institutional economic 19; of meat 2–3, 137, 142
Germany 49, 85, 117, 141
global warming potential 69
Goh Chui Lai 81
Goldman Sachs 66
governance 8, 18, 20, 26, 33–4, 40, 131; spatial 7
government 138; Australian 99, 102–3; British 88, 97; Chinese 5, 17–18, 143; local 20–1, 24, 108; Malaysian 36–7, 40, 81; national 19, 82; New Zealand 76, 79; planning agencies 4; Polish 25; regional 40; Taiwanese 120; US 47, 58, 75
Government Accountability Office (USA) 97–8
governmentality 4, 7–8, 10, 12, 19–20, 23, 26, 32, 34, 44, 61, 108, 131, 136, 142; of food animals 63; of livestock 78–9; of meat 18, 107, 135; of pig industry 30, 33
Grandin, Temple 56–7, 85, 99, 101, 104, 106
grazing land 68
Greek poultry subcontracting 23
greenhouse gas 77; emissions 64, 68–9, 76, 78–80, 82
Greenpeace 64
Groupe Grimaud 43

halal food 101, 104
Harrison, Ruth 87–8
Hatch Act 49
health 7, 43, 45, 80, 84, 117–24, 126–7, 131, 136, 142; animal 46, 52, 88, 103, 138; child 72; occupational 75; public 41–2, 55, 67–9, 74, 76, 80, 97, 137; risks 12, 123; and safety 21, 29, 80, 82; threats 48; workers' 74
Hendra virus 39
Hendrix Genetics 43
Heritage Farms 43
Hindu(s) 140
Hispanic settlements 71
Hope Alliance *see* Pakatan Harapan
Hormel Foods 6
House of Commons (UK) 97
House of Representatives (USA) 97
Huang Chih-Hsiung 120
Hubbard 43
human footprint 77

human–animal relations 8–10, 88, 131
Humane Farming Association 91, 94–5
humane slaughter 103–6
Humane Slaughter Act 87, 97–8
Humane Society of the United States 72
Hungary 24
Hypor 43

Illumina 43
immigrant(s): workers 71, 75–6
in vitro: fertilisation 61; meat 1, 128–9, 132
India 43, 52, 70; beef exports 66
Indian(s) (American) 94
Indonesia 99; animal welfare in 99, 101–2; livestock exports to 100–2
Indonesian: slaughterhouses (abattoirs) 99, 101–3; officials 101–2
industrial: farming 20, 116; meat 61, 63, 82, 84, 106–8, 110, 116, 134–5; slaughterhouses 56, 83–4
industry: agrifood 16; animal science 4; aquaculture 139; beef 102; breeding 54; catfish 18; dairy 48, 51, 78, 136; fishery 139; food 17, 22; food processing 80; genetics 54; livestock 2–3, 6–7, 12, 14–15, 20–1, 33, 40–2, 44, 48, 50, 52, 60–1, 63–5, 67–8, 70, 76, 79–80, 82, 99, 107–8, 115, 120, 124, 130, 134–5, 137–9, 142; manufacturing 75, 137; meat 3, 6–7, 11–12, 20–1, 23, 44, 63–5, 75, 78, 82, 118, 121, 124, **125**, 126, 134–7, 143; meatpacking 127, 137; pig 18–19, 24–6, 29–30, 32–3, 36–40, 43, 50, 55, 60, 66–7, 74, 82, 95, 109–11, 115; poultry 43, 50, 56, 75, 85, 137; pork 65–6; slaughtering 101; timber 64
institution(s) 4, 7–8, 18–19, 24, 51, 61, 138, 143; developmental 5, 110; education 118; and governance 20; governing 40, 61, 64; government 120; health 123; international 139; public 67, 81; science 49, 51
intellectual property 50
intensified farming 1–5, 61
Intergovernmental Panel on Climate Change 78
International Livestock Research Institute 42
International Organisation for Standardisation 77
International Society for Applied Ethology 89

international trade 3
International Vegetarian Union 116, 118
Iowa Select Farms 22
Islamic: consciousness 38; culling 103;
    law 99–101; slaughter 100–1, 104–5
Israel 48, 52
Italy 129
Iyotte, Eva 91

Japan 6, 48, 52, 128, 141
JBS 6, 44, 66
Jensen, Morten 26
Jewish: religious slaughter 98, 101
Jews 140
Jinghong 110–11, *112–13*, 113
Jinuo (people) 110–11, 116
Jinuo Mountain 110, *111*
Jones, Bidda 101
Joyce, Barnaby 102
justice: for animals 10, 142; social 22, 46

Kansas City Federal Reserve Bank 48
Karsy 29
Katter, Bob 102
Kennedy, Robert 91
Kenya 21
Kloppenburg, Jack R. 42, 46–7, 61
Koch, Robert 49
Koffeman, Niko 132
kosher slaughter 101, 104
Krisnamurthi, Bayu 102
Krogman, Todd 92
Kuomintang 120

labour 23, 49, 119, 134; geographies 7;
    immigrant 71; plantation 110; wage
    47; welfare 57, 137–8
Lakota 91–2
land rights 34–6, 38
Landrace pig 71, 110
Laos 71
Latin America 43
Legislative Yuan 120
Lévi-Strauss, Claude 135
life-cycle assessment 77–9
LiveCorp *see* Australian Livestock
    Export Corporation
livestock: commodity production 53;
    farming 1, 3–4, 13, 106, 108, 110, 115,
    119, 134, 138; industry 2–3, 6–7, 12,
    14–15, 20–1, 33, 40–2, 44, 48, 50, 52,
    60–1, 63–5, 67–8, 70, 76, 79–80, 82,
    99, 107–8, 115, 120, 124, 130, 134–5,
    137–9, 142; political economy of 3, 7,

20, 40, 107, 134; production 2, 11, 14,
    40–2, 44, 63, 67–70, 76, 79, 82–3, 108;
    subcontracting 20–3, 40
Livestock and Dairy Development
    Department (Pakistan) 103
Livestock's long shadow (FAO) 64, 119
Loos, Trent 92
Lorenz, Konrad 89
Ludwig, Joe 101

Macedonian 76
Malay(s) 35–9; -Muslim 33, 39; political
    party 33; press 39; royalty 39
Malaysia 14, 33–4, *34–35*, 36–7, 40, 81;
    culture 32, 36, 40; East 22–3; pig
    farming in 4, 81
Malaysian Pork Sellers' Association 81
Malaysian(s) 33; government 36–7, 40,
    81; pig industry 18, 33, 35–9, 82, 111,
    135; states 38
manufacturing industry 75, 137
Marfrig 6
market economy 5–6, 25, 44, 66
Marquess Cattle Company 128
Marx, Karl 10, 45, 47
Marxist critique 5, 42
mass production 67, 135
Mazowieckie 29
McDonald's 76, 95
meat: analogues 107–8, 117, 122, 128,
    132–3, 138, 143; companies 19, 22, 24,
    26, 44, 65–7, 136; consumers 6, 10, 67,
    115–18, 120–3, 127–8, 131, 133, 136;
    geographies of 2–3, 137, 142; industry
    3, 6–7, 11–12, 20–1, 23, 44, 63–5, 75,
    78, 82, 118, 121, 124, **125**, 126, 134–7,
    143; organic 107, 115, 126–8, 135;
    political economy of 4, 6–7, 12;
    production 2, 5–6, 11–12, 17–18, 20,
    22–3, 44, 65, 107–8, 115, 118–19,
    121, 130–1, 133, 135, 137–8, 140;
    synthetic 1, 88, 106–8, 117, 128–33;
    trade 26, 102
Meat and Livestock Australia 99
Meatless Monday 72, 78, 136; Platform
    121–3
Meatout 117
meatpacking 75–6; companies 6, 21, 44,
    65–6, 131, 137; industry 127, 137;
    workers 75; in USA 6, 44, 66
Mednansky, Oleta 91–2
Mekong Delta 138
Melaka Pig Enactment Act 38
Mexico 21, 51, 66, 71

Ministry of Agriculture (China) 21, 66
Ministry of Agriculture (France) 56
Ministry of Agriculture (Malaysia) 36, 81
Ministry of Education (Taiwan) 120
Ministry of Health (Singapore) 122
Modern Meadow 129
Mokrzk 29
Monsanto 47–8
Morliny 26
Morrill Acts 49
Mosaic 66
multinational companies 15, 52
multiple ovulation and embryo
  transfer 54
Muslim(s) 38, 140; Malay- 33, 35, 37, 39;
  religious slaughter 98, 103

national governments 19, 82
National Party (Australia) 102
National Renderers Association 58
National Science Foundation (USA) 77
Native American 91
Nazi 83, 141
Neola Spotted Tail 91
neoliberal political economy 5
Netherlands 1, 24, 43, 73–4, 88
New Straits Times 81
New Tech Solutions 56
New Zealand 48, 105; cloning ban in
  128; fisheries 139; livestock system in
  79; meatpacking in 76
Newsham 43
Nipah virus 38–9, 82, 137
Nippon Meat Packers 6
non-governmental organisations 8, 82,
  122, 134, 143
North America 41, 51; pig industry in 32
North American: companies 44;
  insemination companies 42; meat
  industry 137; *see also* American
Northern Territory Cattlemen's
  Association 102
Norway 66

Office for Vertical Integration of
  Agriculture (China) 21
Office International des Epizooties 51,
  74, 91, 99–100; *see also* Organisation
  Mondiale de la Santé Animale
organic: dairy 62; farming 18, 107–10,
  133; food 10, 16; livestock farming 106,
  108, 119, 110; meat 107, 115, 126–8,
  135; milk producers 48; pig farming 43,
  108, 115; slaughterhouse 56

Organisation Mondiale de la Santé
  Animale 91; *see also* Office
  International des Epizooties
Oryx and Crake (Atwood) 129

Pakatan Harapan 33
Pakatan Rakyat 33
Pakistan 103
Parti Islam Se-Malaysia 33
Pasteur, Louis 49
People for the Ethical Treatment of
  Animals 118, 128
People's Alliance *see* Pakatan Rakyat
Perdue 43
Pew Commission 68
pharmaceutical companies 42
PIC 43
Pig Improvement Company 31
pig(s) 2, 7, 17, 22, 51, 58–62, 66, 69–74,
  78, 84–7, 90, 139, 142; breeding 31,
  43, 50, 59; in Britain 97; in Canada
  73; in China 43, 66, 83, 108–11,
  112–13, 113–6; cloned 128; in
  Denmark 54–5, 71; diseases 74; in
  France 73; genetics 43, 71, 79;
  industry 18–19, 24–6, 29–30, 32–3,
  36–40, 43, 50, 55, 60, 66–7, 74, 82, 95,
  109–11, 115; farm(s) 6, 24–5, 27, 29,
  32–5, *34–35*, 37–9, 64, 66, 71, 81;
  farming 4, 17, 20, 24, 33–9, 43, 108; in
  Malaysia 18, 33, 35–9, 81–2, 111, 135;
  in Netherlands 73; organic 43, 108,
  115; in Poland 4, 18–20, 24–7, *28*,
  29–32, 40; processing 26; in USA 43,
  53–4, 64–5, 67, 70–1, 73, 75, 83, 91–5,
  97; in Vietnam 74
PK Livestock 103
Poland 13, 19–20, 25, 27, *28*, 29–32, 39,
  66; foreign direct investment in 19, 24;
  meat trade in 26; pig farming in 4, 20,
  24, 30; subcontracting in 21
Poldanor 25–7, 29, *29*, 31 –2
Polish: agriculture 25; farmers 30;
  pig industry 4, 18–20, 24–7, *28*,
  29–32, 40
Polish United Workers' Party 25
political: ecology 64, 79; party 33; power
  7; regulation 18
political economy 1, 5, 12, 18–19, 64; of
  animal bodies 9; of aquaculture 138;
  of food 13, 16–17; of food animals 10;
  international 15; of livestock 3, 7, 20,
  40, 107, 134; of meat 4, 6–7, 12;
  neoliberal 5; of science 42

politics 5, 7–8, 11, 13, 16–18, 20, 22, 34, 41, 44–6, 61, 83; cultural 14, 33, 40, 78, 81, 135, 139; of food 18, 41; Malaysian 33; personal 117; racial 33; of religion 40; scientisation of 89; of vegetarianism 117; *see also* biopolitics
pollution 35, 45, 71, 126; air 68–9, 76, 84, 95; environmental 12, 34; waste 39, 84; water 68, 70, 76, 95
porcine stress symptom 55
pork 5, 37, 39, 55, 64, 81, 140; in China 5, 17–8, 66, 109; consumption 18, 35, 65–6, 82, 109; industry 65–6; market 17; production 18, 43, 65–6, 70, 82
Posilac 48
PotashCorp 66
poultry 2, 6, 51, 53–4, 56–9, 71, 73–4, 78, 84–5, 118, 139; in Africa 42; breeding 43, 50; consumption 65; in East Malaysia 22–3; genetics 44, 49; in Georgia 62; industry 43, 50, 56, 75, 85, 137; subcontracting 23; in USA 20, 43, 50, 54, 57, 65, 75, 83, 85, 96
Poultry Science Association 50
poverty 110, 138; alleviation 108, 110–11, 115–16
Prime Food 27
Prisoned chickens, poisoned eggs (Davis) 83
protectionism 14
public health 41–2, 55, 67–9, 74, 76, 80, 97, 137
Purity and danger (Douglas) 140

racial politics 33
racism 33, 86
Redman, Terry 102
Regan, Tom 130
regime theory 13, 15–17
regional governments 40
regulation: political 18
religion: politics of 40
religiosity 40, 119
religious 116; beliefs 119; communities 105; group 38; harmony 33; problems 37; slaughter 98, 101, 103; values 101
Research on National Needs 77
Rockefeller Foundation 51
Romania 66, 71
Rosebud Sioux 91; Reservation 91
Royal Agricultural Society of England 49
Royal Society for the Prevention of Cruelty of Animals 101
Rubenstein, M. 49

rural: communities 12, 49, 108; culture 67; development 13, 36, 67, 108, 138; employment 25, 27; households 109; production 21; unemployment 27
Russia 45

safety: food 14, 37, 66–7, 74–5, 80, 82, 97–8, 108, 131, 133, 139
Schlosser, Eric 83
Science and Technology Options Assessment 91
scientisation of politics 89
Scotland 85
Seaboard Corporation 72
Seaboard Foods 70–1
Securities and Exchange Commission (USA) 72
seed commodification 46
sheep 51, 59, 63, 84, 97–8, 102–3; in Australia 105; in New Zealand 105; in USA 75, 128
Shuanghui International Holdings *see* WH Group
Sičháŋǧu Oyáte *see* Rosebud Sioux
Sim Ah Hock 81
Singapore vegetarianism in 120, 122, 124, **125**
Singaporeans 122
Sinochem 66
Slaughterhouse (Eisnitz) 83
slaughterhouse(s) 58–61, 67, 70, 85, 95, 97–8, 106; in Australia 76, 105; beef 75; in Britain 97; Chicago 55; Indonesian 99, 101–3; industrial 56, 83–4; Malaysian 37; organic 56–7; Polish 25; poultry 75; Turkish 98; United States 60–2, 65, 75, 98; workers 72, 76, 83, 87, 97–101, 105
slaughtering: industry 101; religious 98, 101, 103
small-scale: farmers 12, 21, 49, 60; farming 36; livestock production 108
Smithfield Foods 6, 17, 26, 31, 43–4, 54, 65–6
Smithfield Poland *see* Agri Plus
Smithfield Premium Genetics 43
social: justice 22, 46; movements 41; norms 7, 19
socialism 25
Society for Veterinary Ethology *see* International Society for Applied Ethology
South Africa 117
South America 68

South Dakota Magazine 95
South Korea 141
Southeast Asia 18, 33
Soviet Union 25
Spain 66
speciesism 86
state capitalism 45
state-owned farms 25–6, 32
states: Malaysian 38
Stoczkiewicz, Magda 135
stunning 56, 96, 98–9, 101, 104–6
subcontracting 20, 45; livestock 20–3, 40; poultry 23
Sun Prairie *see* Bell Farms Group
Sweden 85
Swift & Company 66
Swine Influenza Virus Surveillance Program 74
synthetic: hormones 47, 70; meat 1, 88, 106–8, 117, 128–33

Taiwan 58, 120–1, 124, **125**
Taiwanese 121–3, 126, 132; activists 122, 124
technological change: in livestock industry 15, 130
technology 2–3, 40, 45, 47, 54, 61, 64–5, 82, 107, 128, 124; animal science 6; of livestock production 41; production 4, 20, 46, 49, 78
Thailand 51, 83
thanatopolitics 83, 86
Thiel, Peter 129
timber industry 64
Tinbergen, Nikolaas 89
Topigs 43
Totemism (Lévi-Strauss) 135
trade 45, 51, 80; Chinese 18; international 3; in live animals 98, 101–2, 106; livestock 15; meat 26, 102
transnational: capital 139; companies 14; food production 14; investment 19–21, 44
Treasury Cattle Commission (USA) 50
Turkish: slaughterhouse 98
Turostowo 25, 27, *28*, 31
Tyson Foods 6, 43, 65

Ukraine 71
undocumented workers 75
United Food and Commercial Workers International Union 98
United Malays National Organisation 33

United Nations 78, 120
United Nations Development Programme 37
United Nations Food and Agriculture Organisation 42, 64, 68, 76, 83, 91, 116, 119
United Nations Framework Convention on Climate Change 79
United States 26, 42–3, 46–7, 49, 73, 95; ag-ag in 72; agricultural policy 14; animal health in 52; animal science in 42, 51–2; artificial insemination in 51; beef cattle in 54; cloning in 128; dairy industry in 52, 85; farms 22; food retailers in 18; food safety in 97–8; groundwater depletion in 70; livestock industry in 68, 80; meat consumption in 133; meat production in 65; meatpacking companies in 6, 44, 66; pig industry in 43, 53–4, 64–5, 67, 70–1, 73, 75, 83, 91–5, 97; poultry industry in 20, 43, 50, 54, 57, 65, 75, 83, 85, 96; sheep in 75, 128; slaughterhouse(s) 60–2, 65, 75, 98; vegetarianism in 117; workers 75
Upjohn 47
urbanisation 15

ValAdCo 95
vegan(s) 117, 121, 123
Vegetarian Butcher 132
Vegetarian Society (UK) 116
Vegetarian Society Singapore 122–4
vegetarianism 88, 107–8, 116, 118–24, **125**, 127, 133; activists 117, 119, 123, 126, 133; advocacy 107, 116–18, 127–8, 132; politics of 117; in Singapore 120, 122, 124, **125**; in USA 117
Veterinary Division (USA) 50
ViaGen 128
Victorian Farmers Federation Egg Group 57
Vietnam 15, 74, 83, 138, 141; pig industry 74; workers from 71
Vietnamese: catfish industries 18
Vion 6

Wall's Meat Co. 62
Washington Post 83
water: consumption 76, 91; contamination 91; pollution 68, 70, 76, 95; quality 70; resources 68; use 70
Waterkeeper's Alliance 91

Welfare Quality® project 91
Wellard 103
Western European: meat companies 44
WH Group 17, 44
White, Lyn 99, 101
Whole Foods 133
Wielkopolska 24–6, 29
Woodstock Farm Sanctuary 61
worker(s) 10, 41, 45, 75–6, 93–9, 103, 133–4, 142; concentrated (confined) animal feed operation(s) 71–4, 137; deaths 38, 73; exploitation 83; factory farm 67; farm 65, 73; feedlot 60; food animal 10, 60; injuries 75, 94, 137; neglect 72; slaughterhouse 72, 76, 83, 87, 97–101, 105; undocumented 75; unskilled 23; welfare 142
World Bank 42, 110, 120
World Conference on Animal Production 52
World Organisation for Animal Health *see* Organisation Mondiale de la Santé Animale
world system paradigm 14
World Vegetarian Congress 118
Wyatt, Dean 97

Xishuangbanna 110, *111*

Yara 66